Math and Bio 2010

Linking Undergraduate Disciplines

Math and Bio 2010

Linking Undergraduate Disciplines

edited by

Lynn Arthur Steen
St. Olaf College

Published and Distributed by
The Mathematical Association of America

"Meeting the Challenges" Organizing Committee

MARIA ALVAREZ, Professor of Biology, El Paso Community College

CELESTE CARTER, Program Director, Biology, National Science Foundation

CARLOS CASTILLO-CHAVEZ, Biometrics Unit, Cornell University

AMY CHANG, Education Department, American Society for Microbiology

DENNIS DETURCK, Professor of Mathematics, University of Pennsylvania

IRENE ECKSTRAND, Program Director, National Institute of General Medical Sciences

LEAH EDELSTEIN-KESHET, Professor of Mathematics, University of British Columbia

ROSCOE GILES, Deputy Director, Center for Computational Science, Boston University

JULIUS JACKSON, Professor of Microbiology, Michigan State University

JOHN JUNGCK, Professor of Biology, Beloit College (Committee chair)

MADELEINE J. LONG, American Association for the Advancement of Science

DANIEL MAKI, Professor of Mathematics, Indiana University

SHIRLEY MALCOM, Director of Education and Human Resources, AAAS

MICHAEL PEARSON, Associate Executive Director, Mathematical Association of America

LYNN ARTHUR STEEN, Professor of Mathematics, St. Olaf College

TINA STRALEY, Executive Director, Mathematical Association of America

PAUL TYMANN, Associate Professor of Computer Science, Rochester Institute of Technology

ROBERT YUAN, Professor of Cell Biology and Molecular Genetics, University of Maryland

Meeting the Challenges

A Joint Project of

ADVANCING SCIENCE, SERVING SOCIETY

www.aaas.org

AMERICAN
SOCIETY FOR
MICROBIOLOGY

www.asm.org

www.maa.org

Funding for Meeting the Challenges Provided by

National Science Foundation Division of Undergraduate Education,
and the National Institute of General Medical Sciences,
National Institutes of Health, Department of Health and Human Services,
through NSF Grant No. DUE-0232823

www.nsf.gov

www.nigms.nih.gov

www.nih.gov

www.os.dhhs.gov

Additional Material for this Report Provided by

Project
Kaleidoscope

www.pkal.org

The
Reinvention
Center
AT STONY BROOK

www.sunysb.edu/Reinventioncenter/

Foreword

Math and Bio 2010 grew out of "Meeting the Challenges: Education across the Biological, Mathematical and Computer Sciences," a joint project of the Mathematical Association of America (MAA), the National Science Foundation Division of Undergraduate Education (NSF DUE), the National Institute of General Medical Sciences (NIGMS), the American Association for the Advancement of Science (AAAS), and the American Society for Microbiology (ASM). A kick-off meeting, funded by NSF DUE and NIGMS, was held on February 27–March 1, 2003, in Bethesda, MD, and brought together representatives from the various disciplines with an interest in strengthening interdisciplinary efforts in undergraduate mathematics and biology education. Several of the articles in this volume were written by discussion leaders at that meeting, and those articles are informed by the variety of viewpoints and experiences that participants at the meeting shared during their time together.

In addition to the discussion sessions, we heard formal presentations on the challenges facing undergraduate educators in this area. Michael Summers, Professor of Chemistry and Biochemistry, University of Maryland Baltimore County, and Howard Hughes Medical Institute Investigator, spoke on his institution's program that is designed to attract and retain students from groups that are underrepresented in the mathematical sciences and to engage these students through its inclusion of problems in biology that require a quantitative approach. Mary E. Clutter, Assistant Director, Biological Sciences, National Science Foundation, shared her perspective on what the biological sciences will look like over the new century. James Cassatt, Director, Division of Cell Biology & Biophysics, National Institute of General Medical Sciences, National Institutes of Health, shared his views on the importance of increasing work in this area. Judith Ramaley, Assistant Director, Education and Human Resources, National Science Foundation, spoke of the need to break down disciplinary boundaries and called on us to seek ways to educate students to adapt to a changing world. Lou Gross, Professor of Ecology and Evolutionary Biology and Mathematics and Director, Institute for Environmental Modeling, University of Tennessee, Knoxville, discussed his experiences developing new courses and the need to enhance quantitative training for all biology students, and to motivate them to understand its importance in modern biology. Abdul-Aziz Yakuba, Howard University, and Carlos Castillo-Chavez, Arizona State University, shared their views on current trends in the integration of biology, computer science, mathematics and statistics, and how those trends are driving the need to reconsider the way we train our students.

Additional information on most of the presentations mentioned above can be found on the project website, www.maa.org/mtc. A number of articles, reports and links to external resources are collected on the site to support faculty efforts to develop or enhance interdisciplinary education. We anticipate maintaining and expanding this site over the next two years.

When we began to prepare this report, we were offered opportunities to collaborate with other groups who were interested in pursuing similar questions. Wendy Katkin, director of the Reinvention Center at the State University of New York at Stony Brook, directed a survey aimed at determining the current state of interdisciplinary work that is being carried out at research institutions, in particular the impact such work has on undergraduate training for students in the biological, mathematical and computer sciences.

An overview of their findings is included in this report. We were also fortunate in developing a connection to a group of Project Kaleidoscope's Faculty for the 21st Century (F21). Through a survey of this group, a number of courses and special initiatives that have or are being developed to enhance cross-disciplinary training for undergraduates in these disciplines were identified. Short descriptions of these efforts are included in this report. We owe special thanks to Debra Hydorn, Department of Mathematics at University of Mary Washington, who played a lead role in collecting and organizing this material. Stokes Baker and Jeffrey Boats, both at the University of Detroit Mercy, also helped gather and prepare the program profiles you'll find here. In addition, Project Kaleidoscope director Jeanne Narum was a tremendous help in bringing about this study.

Project leaders are pursuing a variety of outreach activities designed to stimulate discussion of strategies for meeting the goals identified through this project. A panel discussion, organized by Calvin Williams of NSF DUE, was held at the 2003 Joint Mathematics Meetings and helped spur interest while we were still in the planning phase for Meeting the Challenges. An article on the project, written by Victor Katz, appeared in FOCUS, the newsletter of the MAA, following the kick-off meeting. John Jungck organized a short course, *Reading the Book of Life: How Bioinformatics Makes Sense of Molecular Messages*, prior to MathFest 2003, the MAA's annual summer meeting (Boulder, CO). Contributed paper sessions focused on mathematics and biology, organized by the MAA Committee on Mathematics across the Disciplines, were held during MathFest 2004 (Providence, RI). A second session is planned for the 2005 Joint Mathematics Meetings (Atlanta, January 5–8). A symposium, chaired by John Jungck, is planned for the 2005 AAAS Annual Meeting (Washington, February 17–21). Through the MAA's Professional Enhancement Program (PREP), workshops on mathematics and biology have been offered during both summer 2003 and 2004. While at this writing the 2005 PREP schedule has not been set, it is likely that additional workshops to support the broad goals of Meeting the Challenges will be offered next year. (The 2005 PREP program announcement will be available on the project website, `www.maa.org/prep`, in late fall 2004.)

Funding for the entire project, including this report, was provided by NSF-DUE and NIGMS. Calvin Williams and Lee Zia at NSF DUE and Irene Eckstrand at NIGMS provided encouragement and advice throughout the planning stages, and helped us identify and recruit both speakers and participants for our kick-off meeting last year. They were also supported by their colleagues, many of whom joined in our discussions and thus contributed to the articles you find here. We are grateful for all of their help.

As we were planning this project in fall 2002, the National Research Council released *Bio 2010: Transforming Undergraduate Education for Future Research Biologists*. This report presents a broad call for reconsidering the presentation of engineering and computer science, chemistry, physics and mathematics to life science students. While we did not anticipate this call, our report can be seen as a partial response to the questions raised in *Bio 2010*. Reports from other groups have called for consideration of similar questions, including the reports on the relationship between mathematics and the biological sciences contained in the report of the *Curriculum Foundations Project: Voices of the Partner Disciplines*, sponsored by the MAA Committee on Curriculum Reform and the First Two Years. We certainly hope that the ideas contained here will be viewed as a useful and constructive contribution to this ongoing dialog.

Michael Pearson
Mathematical Association of America

Contents

Introduction

Math and Bio 2010 envisages a new educational paradigm in which the disciplines of mathematics and biology, curricularly quite separate, will be productively linked in the undergraduate science programs of the twenty-first century. As a science, biology depends increasingly on data, algorithms, and models; in virtually every respect, it is becoming more quantitative, more computational, and more mathematical. While these trends are related, they are not the same; they represent, rather, three different perspectives on what many are calling the "new biology."

All three methods—quantitative, computational, mathematical—are spreading across the entire landscape of biological science, from molecular to cellular, organismic, and ecological. The aim of this volume is to alert members of both communities—biological and mathematical—to the expanding and exciting challenges of interdisciplinary work in these fields. Ten chapters, summarized below, outline issues, examples, resources, and challenges. Appendices provide references to resources—people, textbooks, monographs, web pages—to encourage exploration and help faculty realize the vision of *Math and Bio 2010*.

The educational challenges facing the new biology stem from many sources including students' weak mathematical preparation, faculty's mono-disciplinary training, and departments' tradition of curricular silos. In "Challenges, Connections, Complexities" (pp. 1–12), Beloit College biologist John Jungck uses recent resources and reform recommendations to outline a strategy to educate biologists, mathematicians, and computer scientists for what he calls "complexity"— either research problems that are scientifically and mathematically complex or educational challenges that are pedagogically and curricularly complex. Jungck provides pointers to helpful resources, but also notes continuing difficulties created by publishers, departments, and faculty who too often retreat from the mathematical tools that students clearly need.

Although biology has historically been the least mathematical of all the sciences, it has never been divorced from mathematics. Many major parts of biology (notably evolution, ecology, and physiology) have benefited from mathematical models. Nonetheless, because biological problems are so complex, the fit between theory and data has never been as good in biology as in the physical sciences. In "The 'Gift' of Mathematics in the Era of Biology" (pp. 13–25) St. Olaf College mathematician Lynn Steen traces the history of mathematical models in biology since Malthus' famous essay on the "principle of population." Only now, two centuries later, has the power of mathematical tools caught up with the complexity of biological problems so that biology, like physics, can speak clearly in the language of mathematics.

A major catalyst for "Meeting the Challenges" was publication of *Bio 2010: Transforming Undergraduate Education for Future Research Biologists*. This report from the National Research Council (NRC) set forth two central propositions: first, biology has become interdisciplinary, and second, this change requires fundamental restructuring of biology education. In "A Quantitative Approach to the Biology Curriculum" (pp. 27–33), University of Maryland cell biologist Robert Yuan, a senior NRC consultant on the *Bio2010* project, focuses this argument on the implications of a more quantitative biology curriculum. Students, he asserts, will need better and different quantitative preparation—but first faculty need to clarify the quantitative needs of the new biology. Courses need to be created and re-created, both in biology and in mathematics. Faculty—likewise in both biology and mathematics—need incentives, time, and resources to learn new areas, to develop new networks, and to reconnect teaching with research. Finally, colleges and universities need to change a culture that favors research over teaching and a structure that binds departments to disciplines. Yuan calls these "issues to consider." In fact they are much more than issues: they are imperatives.

Much has been made of the DNA-inspired transformation of biology into an information science where genes and proteins play the role of "wetware" in a computational view of living organisms. Less recognized is a parallel revolution brought about by the development of computationally intensive tools for *visualizing* biological phenomena. The aphorism "a picture is worth a thousand words" reflects the extraordinary power evolution conveyed to the human eye. In "Visualization Techniques in the New Biology" (pp. 35–43), biologist Maria Alvarez of El Paso Community College describes how computational tools greatly enhance the eye's natural power of sight and insight, of recognition and remembrance. Students and researchers can now both see and, as important, measure phenomena that are beyond the range of ordinary optics. Scales of time and space can be stretched and shrunk; interiors sliced and scanned without intrusion; images magnified, colored, and animated; perspectives changed and structures compared. These tools offer not only powerful aids to research, but also motivational opportunities for the many students who will understand from a graph what they fail to grasp from a table of data.

It seems tautological to say that the interdisciplinary character of the "new biology" requires an interdisciplinary education. But for most colleges and universities, an interdisciplinary approach to basic science education is more like a contradiction than a tautology. Interdisciplinary work is the exception, not the rule, and where it does occur it is most often at the end of a major where it has to fight against already fortified disciplinary silos. In "Building the Renaissance Team" (pp. 45–50), University of Montana biologist Carol Brewer and Indiana University mathematician Daniel Maki argue for a "renaissance campus" focused on students' needs to learn to function in a world without disciplinary boundaries. To this end they suggest that in all courses faculty should highlight relationships with other disciplines more than the distinctiveness of their own fields, and that from their earliest encounter with science, students should work in teams on projects that require cross-disciplinary skills. The benefits of shared knowledge, cross-field communication skills, and effective teamwork will serve students well as they move into graduate and professional positions where such experiences are becoming the norm.

The "race" to decode the human genome put the new biology on the map of public awareness. Only then did people outside the field become really aware of how dependent this part of biology is on high-speed computers. Inside the academy, this recognition produced a new field with the European-sounding name of *bioinformatics*. Because of this history, bioinformatics is often thought of as the science behind the decoding of genomes—that is, the creation and deployment of algorithms that stitch together the molecular codes for fragments of DNA or RNA. In his essay "Bioinformatics and Genomics" (pp. 51– 61), Julius Jackson of Michigan State University argues for a more expansive view of bioinformatics as dealing with the structure and dynamics of information in any biological system. From this perspective, bioinformatics is a synthesizing science that helps transcend deficiencies that have become apparent in the last century during which the paradigm of biological research was to reduce complex processes to finer and finer constituent pieces. To this end, Jackson argues for educational structures much like the renaissance campus envisioned by Brewer and Maki where inter-, cross- and multi-disciplinarity is the rule, not the exception.

Notwithstanding the trendy appeal of bioinformatics, the quantitative needs of the new biology extend broadly throughout the entire field. All areas of mathematical modeling in biology have been enhanced with improved digital instrumentation (ranging from satellite imaging and medical monitors to transmitters attached to wild animals). Computer networks enable this data to be pooled and compared, for the first time on a worldwide basis. In "Adapting Mathematics to the New Biology" (pp. 63–73), mathematician Leah Edelstein-Keshet of the University of British Columbia describes various approaches to making undergraduate mathematics courses—especially entry level courses—more suitable for the enhanced mathematical needs of biology students. These approaches, which Keshet illustrates with examples of actual courses, include different emphasis on subject matter, enhanced use of biological models and real data, and support systems to help math-averse students make the transition to the more quantitative new biology. All these, and more, can be enormously enhanced through "positive, early research experiences."

Unlike mathematics which has a long history of engagement with biology, computer science is a new field whose major connection with biology is manifest in the products of bioinformatics (genetic maps), visualization (computer-enhanced pictures), and networked databases. Beyond these instrumental uses of computers, however, are tantalizing hints of even more fundamental connections based on the common theme of information. Insights from genomics suggest that life—what biologists study— is as much an information process as a physical process. And the study of information processes is the subject of computer science. Diverse possibilities of this alignment are suggested in the report "Computer Science and Bioinformatics" (pp. 75–82) that was prepared by an informal group of computer scientists who have been working at this interface. They note both practical and theoretical characteristics of algorithms (e.g., computational intractability) that have significant implications for biology, as well as examples of biological processes that pose new challenges for computer science. Here too, a major challenge is how to integrate two disciplines with very different traditions and histories in the experiences of undergraduates.

Fueled with grants from the National Institutes of Health (NIH) and the Howard Hughes Medical Institute (HHMI), the biological sciences have become significant centers of research activity at major universities across the country. As noted above (and throughout this volume), much of this research rests on new mathematical or computational foundations. To gauge the degree to which innovation and excitement from the research frontier in biology is reflected in undergraduate programs, the Reinvention Center at the State University of New York at Stony Brook undertook a study that asked particularly about recent undergraduate curricular and structural change, especially at the intersection of biology, mathematics, and computer science. "Building Connections in Research Universities" (pp. 83–99) reports the results of this survey. It includes quantitative indicators of the extent of change, examples of new courses or programs, and extensive first-hand comments by faculty about their motivations and accomplishments as well as the challenges and impediments they faced.

Historically, much of the innovation in college-level science education has come not from research universities but from liberal arts colleges whose primary focus is undergraduate education. So to complement the Reinvention Center's study of research universities, Project Kaleidoscope undertook a parallel investigation of liberal arts colleges and non-doctoral universities. The resulting report, "Quantitative Initiatives in College Biology" (pp. 101–119) profiles nearly twenty programs at different institutions. Some are new or revised courses, others are programs involving several related courses. Some center in biology, some in mathematics, some in computer science. Some focus on courses for entering students, others for students in the middle or end of their college studies. A few are well on their way to becoming truly joint programs of the kinds envisioned by other authors in this volume.

Acknowledgements. Impetus for this report came from a working conference of mathematical and biological scientists hosted by the Mathematical Association of America, the American Association for the Advancement of Science, and the American Society for Microbiology with support from the National Science Foundation and the National Institutes of General Medical Sciences. Leadership for the project was provided by an Organizing Committee (p. iv) led by John Jungck, professor of biology at Beloit College. The editor and project director wish to thank all those who helped bring about this linking of disciplines.

Lynn Arthur Steen
St. Olaf College
July 2004

Challenges, Connection, Complexities: Educating for Collaboration*

John R. Jungck
Beloit College

The avalanche of data generated by contemporary high-throughput biology has challenged traditional methods of drawing biological inferences, primarily because inferences based on these data require simultaneous consideration of multiple variables, multi-dimensional visualization, and often also multi-disciplinary analyses. Two disciplines that have extraordinary potential to help address these biological challenges are mathematics and computer science. Furthermore, mathematics will be necessary to the development of succinct summaries of the causal mechanisms underlying such complex phenomena.

While mathematics has played exceptionally important roles throughout the history of biology, too frequently it has been underrepresented in biology curricula because textbook authors—as well as many professors—assume that biology students have inadequate mathematical preparation. But in recent years, computer science and mathematics have completely transformed the practice of biology. Thus the absence of strong curricular ties between the biological and quantitative sciences *misrepresents contemporary biological research,* deskills many biology, computer science, and mathematics students, and fails to prepare them to collaborate on significant problems.

Fortunately, recent reforms in teaching biology by groups such as the BioQUEST Curriculum Consortium,[1] in teaching calculus by groups such as Project CALC,[2] in teaching computational science by groups such as the National Computational Science Institute (NCSI),[3] in teaching chemistry by groups such as ChemLinks,[4] and in teaching physics by groups such as Workshop Physics,[5] have empowered thousands of American undergraduates. These students have become proficient in the use of software packages that can be used to investigate the behavior of many famous mathematical models in biology, collect and mine complex data sets, and evaluate their hypotheses with multidimensional visualizations.

To make good use of these many software packages in biology, students need important and effective mathe-

John Jungck is Mead Chair of the Sciences and professor of biology at Beloit College in Wisconsin. A specialist in mathematical molecular evolution and science education, Jungck is co-founder of BioQUEST, a national consortium of college and university biology educators devoted to curricular reform. Jungck is a contributor to the National Research Council report *Bio 2010* and a Fulbright Scholar at Chiang Mai University in Thailand where he offers workshops on computational molecular biology and molecular bioinformatics. A graduate of the University of Minnesota, Jungck received his Ph.D. in 1973 from the University of Miami.

* This article is based upon work supported by the National Science Foundation (NSF) under Grant No. 0127498. Any opinions, findings, and conclusions or recommendations expressed are those of the author and do not necessarily reflect the views of the NSF.

[1] www.bioquest.org/

[2] www.math.duke.edu/education/proj_calc/

[3] www.computationalscience.org

[4] chemlinks.beloit.edu/

[5] physics.dickinson.edu/~wp_web/WP_homepage.html

I

matical models. Here too we find a plethora of relatively new resources. First, each of the curricular reform initiatives listed above has generated a great deal of related relevant material that includes simulation, modeling, data mining, image analysis, problem solving, and multidimensional visualization. Second, numerous recent texts in mathematical biology, research journals, web sites, and some advanced biological texts are replete with quantitative models of all kinds. Third, a variety of college and university biology education journals provide peer-reviewed examples to consider. Fourth, professional societies in mathematics, biology, and computer science have published many articles and monographs that demonstrate effective applications of mathematical models to biological phenomena.

Recent recommendations for reform of undergraduate science and mathematics education have reinforced the need for more mathematics and computer science in undergraduate biology education as well as more attention to biological applications in mathematics and computer science education. When these recommendations are examined in light of current research and available resources, one finds many progressive resources for those who seek to develop programs that address these challenges, that make necessary connections, and that handle the complexities of collaboration among biologists, mathematicians, and computer scientists.

Challenges: Merging Research, Teaching, and Learning

Recommendations for closer alliance between mathematics and biology have come from several sources. Most notable, perhaps, is the *Bio 2010* report from the National Research Council (NRC, 2003). Both of its first two recommendations (excerpted below) for the reform of undergraduate education in the life sciences emphasize mathematics:

> **Recommendation 1:** Given the profound changes in the nature of biology and how biological research is performed and communicated, each institution of higher education should reexamine its current courses and teaching approaches to see if they meet the needs of today's undergraduate biology students. Those selecting the new approaches should consider the importance of building a strong foundation in mathematics and the physical and information sciences to prepare students for research that is increasingly interdisciplinary in character.

> **Recommendation 2:** Concepts, examples, and techniques from mathematics, and the physical and information sciences should be included in biology courses, and biological concepts and examples should be included in

other science courses. Faculty in biology, mathematics, and physical sciences must work collaboratively to find ways of integrating mathematics and physical sciences into life science courses as well as providing avenues for incorporating life science examples that reflect the emerging nature of the discipline into courses taught in mathematics and physical sciences.

In a similar vein, the Mathematical Association of America (MAA) has released several major curricular reports that explicitly address the interactions between biology and mathematics. Four reports stand out as particularly relevant. Three of these, emanating from a series of workshops organized by MAA's committee on Curriculum Renewal Across the First Two Years (CRAFTY), have been bound together with several other reports in a single report (Ganter & Barker, 2003):

- *Biology* (Organized by David Bressoud and edited by Judy Dilts and Anita Salem);
- *Health-Related Life Sciences* (Organized by Reuben W. Farley and William E. Haver and edited by Thomas F. Huff and William J. Terrell);
- *Technical Mathematics: Biotechnology and Environmental Technology* (Organized by Bruce Yoshiwara and Gwen Turbeville and edited by Elaine Johnson, John C. Peterson, and Kathy Yoshiwara).

The fourth report is a major publication from MAA's Committee on the Undergraduate Program in Mathematics (CUPM) that recommends goals and strategies for undergraduate mathematics in the first decade of the twenty first century (CUPM, 2004). All four MAA reports document enormous changes that are currently occurring in undergraduate mathematics courses, impacting not only mathematics majors but also students across science and engineering.

The Society for Mathematical Biology[6] sponsors an annual symposium on mathematical biology education. In recent years five such meetings were held—two at the University of Tennessee-Knoxville, one each at the University of Mexico, Iowa State University, and Beloit College. Lou Gross, president of the society, maintains a web site[7] on which he has archived many files related to these five meetings. These files and related software will be of considerable interest to curriculum developers and departments seeking already-developed materials.

In 1998 Pomona College hosted a conference on computing in the life sciences[8] at which leaders identified

---

[6] www.smb.org

[7] www.tiem.utk.edu/~gross/

[8] www.nsf.gov/bio/bioac/workshop/pomona.htm

major areas in which computers had significantly transformed the relationship of teaching to research.

- Real-time data acquisition has replaced oscilloscopes, strip chart recorders, and kymographs. (Costs have dropped considerably since 1998 and equipment has become much smaller and more portable.)
- Digital video microscopy and scanners are ubiquitous in the lab and camcorders in the field.
- Quantitative image analysis and morphometry software are freely available.
- Massive online resources of biological data such as the digital human female and male cadavers and extensive bioinformatics databases are easily accessible.
- Geographic information systems (GIS) and global positioning systems (GPS) have become easily portable for field work (and with the deregulation of satellite data, much more accurate).
- Optical sectioning allows generation of three-dimensional reconstructions of cells, organs, tissues, and organisms. (Recently, 3D projection systems such as GeoWall[9] have tremendously increased the classroom use of 3D visualizations.)
- Simulation has replaced experiments and field studies that would be too dangerous, expensive, or lengthy to perform and study in a laboratory or field context.
- Areas of biology wholly reliant on computational power have exploded in importance: bioinformatics; X-ray crystallography of the 3D structures of biological macromolecules; construction and analysis of phylogenetic trees; laser confocal microscopy, molecular visualization of ligand docking; combinatorial chemistry; subtraction of Brownian motion to visualize beyond the limits of conventional optical microscopy; prototyping experimental designs with simulated conditions; and data mining (then called KDI for Knowledge and Distributed Intelligence).
- Multimedia capabilities have enabled biologists to replace 35mm slides with PowerPoint presentations. These technologies have been widely adopted in recent years with visually rich materials such as animated "gif" files or QuickTime movies now posted on numerous lab websites.
- 'Worldware' like e-mail and list-serves have changed not only how scientists communicate with one another,

but also how students and faculty communicate outside of class.

Of course, this list is the product of the particular experiences and interests of those who assembled in 1998 and could easily be revised and expanded by a new group charged with the same task today. Nonetheless, the widespread presence of items on this list have greatly influenced research and education in physiology, ecology, neurobiology, biochemistry, cellular biology, molecular biology, anatomy, developmental biology, and biophysics. Of course, these efforts have been tremendously enhanced by the use of computers, often purchased through Instructional Laboratory Instrumentation (ILI) grants from the Division of Undergraduate Education (DUE) at the National Science Foundation.

At the Pomona meeting, Diane Card Linden of Occidental College emphasized that "open-ended experiments that students could 'own' increased excitement they have about neurobiology." Newton Copp of Claremont McKenna College described his Human Physiology course as being able to focus more on learning biology, not using equipment; on analyzing, not just collecting data; and on developing students' abilities to ask and answer questions. Copp concluded that "the most positive outcomes have been increased sizes of samples, increased variety and speed of analysis, better graphs, and a general shift of focus to the physiological question being investigated."

Beth Braker of Occidental College noted that "students perceive field biology as a 'non-instrumented science' using only clipboards and pencils." But by using geographic information systems (GIS), global positioning systems (GPS), image analysis, and data loggers, students were able to conduct field work that helped them understand and analyze such processes as photosynthetic rates and nutrient flows in forest canopies, and predation of pollen grains on a stigma.

Although the 1998 Pomona meeting reflects practices at a few research-oriented liberal arts colleges six years ago, in the interim these practices have spread to almost every college and university undergraduate program.

Polarities

The practice of contemporary biology is extraordinarily data-rich; practitioners must regularly use computers in the acquisition, organization, and modeling of these copious resources. However, the underlying mathematics is often hidden in algorithms embedded in silicon chips within laboratory instruments or in compiled computer

---

[9] geowall.geo.lsa.umich.edu/

code for visualization, modeling, and statistical software. For biologists to have a better idea of what is and is not possible, they need at least to be aware of the assumptions and limitations of the underlying mathematical analyses that are being invisibly performed for them.

Furthermore, unless these mathematical procedures are unpacked and made explicitly visible to instrument and computer users, it will be difficult for them to judge what sorts of mathematics and computer science are most appropriate for specific courses and curricular materials. Some of these are old problems. For example, both pH and absorbance are logarithmic transforms (one of hydrogen ion concentrations, the other of optical transmittance). Nonetheless, many biology students do not recognize a one-unit change as a ten-fold multiplication or a 0.3 unit change as a doubling. Today, instruments' computer chips may be calculating so many complicated functions that even a biologist with an undergraduate major in mathematics or computer science may not be able to fully understand what is being computed. The Computer-Assisted-Tomography (CAT) scanner offers a common example.

In an article on ten equations that changed biology (Jungck, 1997a), I illustrated each of the equations with a piece of software that was associated with its use. Subsequently, as I presented talks of the same title at colleges and universities in the U.S. and abroad, many scientists offered additional equations for consideration in such a "top ten" list. However, few were prepared to give up any already on the list. (The one equation that I have been most tempted to add is the Patterson equation because of a convincing argument presented by X-ray crystallographers when I presented at Massey University in New Zealand. Of course their enthusiasm may have been influenced by national pride since Patterson was from New Zealand.)

General agreement on these ten equations is largely the result of how the list was created: it was based on suggestions from many biologists, applied mathematicians, and historians of science. Half the list highlights significant Nobel Prizes in the life sciences whose results are represented in every general biology text. The other half celebrates concepts that have become part of biological "common sense" but whose mathematical origins have been ignored or suppressed. For example, the importance of maintaining diversity in populations so that they have a better chance of surviving a major environmental change is now common sense, but the partial differential equation form of Sir Ronald A. Fisher's "Fundamental Theorem of Natural Selection" never appears in general biology texts and seldom appears even in undergraduate texts on evolution. The intent of the article was not to argue that these ten equations should become the foundation for mathematics and computer science in biology education, but to alert biology educators to the historical and continuing importance of mathematics and computer science to contemporary biology.

Moving beyond equations, one can inquire into the most important computational and mathematical concepts that should be part of the toolkit of every biologist. My previous experience suggests that such a list will be equally fruitful in catalyzing discussion about what we value. For concepts, however, ten is clearly too small a limit. Based on an informal survey of numerous colleagues and drawing on my own thirty-eight years of teaching quantitative, problem-solving biology courses, and with apology for whatever hubris is implied in identifying any such list, I offer in Figure 1 a "top 25" set of polarities: pairs of mathematical or computational ideas and techniques of importance to biology.

While polar mythologies or binary models have their own conceptual problems, it is the difficulty in distinguishing between members of the pairs that so often obscures understanding. They represent conceptual distinctions that many biology students find difficult to grasp and apply. Equally, and as important, they illustrate common failures of communication between biologists on the one hand and mathematicians and computer scientists on the other.

While I might extol the virtues of fuzzy logic and Eastern embracement of both/and paradoxes, or criticize the limits of Aristotelian binary logic and excluded-middle formulations, I believe that these polar dichotomies help catalyze careful distinctions and raise important curricular questions. Although these polarities are subtle and difficult to define precisely, many of the distinctions embedded in them are so fundamental that an expert in one field would not comprehend that a respected colleague in another discipline may not understand them intuitively.

Obviously, any list is debatable. Many items could be added or deleted depending upon one's discipline or experience. I provide this particular list not to be definitive, but to invite discussion about the conceptual difficulties that students face while learning to be a professional in the growing areas where biology, mathematics, and computer science intersect. They also suggest some fundamental ideas that are important for faculty who collaborate with peers in cognate disciplines.

1. relation	function
2. geometry	topology
3. discrete	continuous
4. algorithm	heuristic
5. univariate	multivariate
6. deterministic	predictive
7. correlation	causation
8. causal mechanism	curve fitting
9. specificity	sensitivity
10. precision	accuracy
11. spreadsheet	database
12. hierarchical database	relational database
13. find, identify	sort, classify
14. pattern	process
15. linear increase	combinatorial explosion
16. overdetermined equations	underdetermined by data
17. provability	computability
18. frequency	time
19. capricious	random, stochastic, ergodic
20. bottom up, local, proximate	top down, global, ultimate
21. scale-free network	random network
22. semantic	syntactic
23. differentiation	integration
24. monotonic	hysteretic, catastrophic
25. cipher	code

Figure 1. Pairs of mathematical or computational ideas of importance in biology

Dealing with polarities is not always easy or comfortable. Initial motivation for reform of calculus in the 1980s did not arise from concern about failure of student learning in calculus classes. Instead, it came from a challenge by computer scientists and discrete mathematicians who argued that this 350-year-old branch of mathematics which was fine for Galilean and Newtonian physics was not as relevant to contemporary biologists and other scientists (Ralston 1981, 2004; Tucker 1995). More generally, the discipline of mathematics is under challenge by the digital computer. In a paper entitled "Will the digital computer transform classical mathematics?" Brian Rothman (2003) cites cases where computers have replaced traditional proofs (e.g., for the Four Color Theorem) and argues further that the issues are not only pragmatic but conceptual as well. Similarly, systems biologists are currently critiquing classical molecular biologists for using excessively reductionistic approaches that focus on individual knock-out experiments that yield yes-no answers rather than seeking to understand global phenotypes based on synergistic interactions of signaling pathways.

As these examples illustrate, conversations about how curricula in biology, mathematics, and computer science can serve one another are likely to encounter highly contentious issues. There is an omnipresent risk that diverging sub-disciplinary perspectives and interdisciplinary

- Experience independent learning and research practices.
- Work in a group or team aimed at understanding the nature of collaboration.
- Present and communicate technical subject matter.
- Organize and plan the completion of a major project.
- Prepare findings for publication.
- Learn, through teaching others, the creative aspects of mathematics.
- Read research papers and participate in seminars.
- Evaluate peer research projects.
- Integrate skills that they have acquired in their major.
- Instill an appreciation of the power of mathematics through its ability to shed light on other areas of human inquiry.

Figure 2. Goals for students in research-active biology classrooms.

rivalries might deflect attention from the important task of building extendable foundations for students. These foundations are not discipline-specific. Steve Deckelman (2003, p. 35) at the University of Wisconsin-Stout has identified ten characteristics of research-active classrooms in mathematical modeling of biological phenomena (Figure 2). Note that no particular principles of biology, mathematics, or computer science appear on this list. Yet too often discussions about curriculum degenerate into lists of content and coverage, which have become the enemy of substantial reform. Thus another important polarity is the difference between content skills and broad goals.

Thus, as we proceed, we need to be conscious of three simultaneous trajectories (see Figure 3): the *scientific*

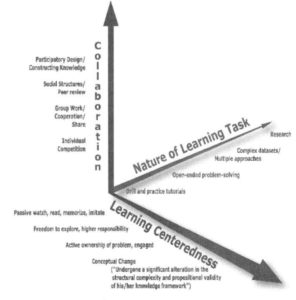

Figure 3. LT²: Learning Through Technology.

(what is the nature of the scientific, mathematical, and computational learning that we expect students to engage in?); the *cognitive* (i.e., what are the conceptual changes in individual knowledge structures and affective relations students demonstrate?); and the *collaborative* (what role do participatory constructions and peer review play in developing membership in a profession?). These "Learning Through Technology" (LT^2) criteria (Jungck et al., 2000) were used in an extensive analysis of the use of technology in undergraduate science, technology, engineering, and mathematics (STEM) education carried out by the National Institute for Science Education at the University of Wisconsin. In this study these criteria were employed to analyze a wide variety of innovations at a diverse set of institutions and in multiple disciplines. (Details can be found under "Case Study Code Definitions" on the "codes" link from the project web site.[10])

Connections: Three Disciplines, One Student

In reporting on a project that examined the status of universal education, demographers David E. Bloom and Joel E. Cohen (2002) identify four different grounds for providing an "education for all": humanitarian, sociological, political, and economic. The need for biologists to have a mathematical and computational education could be based on the same grounds. First, biology students represent the next generation of scientists. They will need to solve both what we have not been able to solve as well as new problems that we have created. Second, since we require biology students to take mathematics and science courses, we need to respect what they have learned by building on that which we have required them to learn. Third, to be credible, we need to act consistent with our own requirements; consider this a sociological warrant of peer review(Jungck, 1997b). Fourth, biology students who learn more mathematics and computer science are more apt to succeed professionally and economically than their peers who have taken less mathematics (Gross, 1994, 2000).

At a 1996 conference on mathematical biology education, Lou Gross of the University of Tennessee-Knoxville (and current president of the Society for Mathematical Biology) concluded:

> It is unrealistic to expect many math faculty to have any strong desire to really learn significant applications of math that students will readily connect to their other

course work, though there is a core group who might do this.

So what do we do to enhance quantitative understanding across disciplines? Here is what I say to life science faculty:

Who can foster change in the quantitative skill of life science students? Only you, the biologists can do this! There are two routes: Either convince the math faculty that they're letting you down, or teach the courses yourself.

Note: Math faculty will not take you seriously unless you show them how the quantitative topics you insist that they cover will be used in your own courses!

This means biology courses must become less a "litany of conclusions," and more an exploration of how and why natural systems came to be as they are.

Unfortunately, this battle must be fought over and over at each institution.

Based on the "Meeting the Challenges" conference, it appears that Gross underestimated the flexibility and interest of mathematics and computer science faculty and overestimated the willingness of biologists. Nearly every mathematician and computer scientist who attended the Challenges conference had already made a commitment to work with biologists, to learn a significant amount of biology, and to serve the majority of their students (who are biology majors of some sort). On the other hand, the majority of biologists at this same conference appeared intimidated by the thought of having to re-learn (or in some cases, learn for the first time) the mathematics and computer science that their biology majors learn in the first two undergraduate years. Most are even less eager to include mathematical and computational examples in their biology courses that depend on the mathematics that students study in the first two years of college. Nonetheless many serious science students arrive at college already having taken and received credit for AP Calculus. So the opportunity exists to collaborate on development of biology curricula that can utilize students' mathematics background or courses that they are simultaneously enrolled in.

Whereas mathematicians have already demonstrated that they can include many significant biological examples in a calculus course (e.g., Neuhauser, 2000; Adler, 1998), there is as yet no first year general biology text with any significant number of equations, quantitative problems, or computational modeling exercises. Quantitatively oriented texts are available for upper-level courses in ecology, genetics, neurobiology, and physiology, but there is a huge gulf between the one or two examples of high school algebra and geometry in a general biology text and the discrete mathematics, statis-

---

[10] www.wcer.wisc.edu/nise/cl1/ilt/case/case.htm

tics, probability, and differential equations used in these upper level biology texts. Who is supposed to provide the mathematical bridge for students?

Those biology educators who have sought to supplement conventional texts and lab books with quantitative problem solving and data analysis have usually relied on problem books for more advanced courses, material from pre-1960 textbooks, and, primarily, the extensive collection of computer software that has been developed for biology education. However, these individuals are the exception rather than the rule. How, then, will we build a community of biologists, mathematicians, computer scientists, and educational researchers capable of addressing the challenges enunciated in *Bio 2010*?

If a student engages a curriculum such as one suggested by *Bio 2010*, each of the three disciplines we are examining (biology, mathematics, computer science) will have an opportunity to interact with students through courses in each area. But *Bio 2010* calls for better interactions with physics, chemistry, and engineering as well. Yet when faculty discuss student difficulties in learning our disciplines, we seldom consider which problems we share in common across disciplines and which issues are specific to our own discipline. Most students experience their curriculum as a series of courses in different disciplines rather than as an integrated interdisciplinary experience. We need to ask what learning we expect to be transferable from one area to another. Are discipline-bound faculty willing (or able) to change curricula in such a way that students will be asked to solve problems that draw on multiple disciplines and to seamlessly integrate multiple perspectives and methodologies to deal with complex problems?

These problems are not new. In "Revolution by Stealth: Redefining University Mathematics," Lynn Arthur Steen (2000) suggests that the "hidden curriculum" has already subverted much of the process:

> But what is somewhat new—and growing rapidly—is the extent to which good mathematics is unobtrusively embedded in routine courses in other subjects. Anywhere spreadsheets are used (which is almost everywhere) mathematics is learned, as it is also in courses that deal with such diverse topics as image processing, environmental policy, and computer-aided manufacturing. From technicians to doctors, from managers to investors, most of the mathematics people use is learned not in a course called mathematics but in the actual practice of their craft. And in today's competitive world where quantitative skills really count, embedded mathematical tools are often as sophisticated as the techniques of more traditional mathematics.

So tertiary mathematics now appears in three forms:

as traditional mathematics courses (both pure and applied) taught primarily in departments of mathematics; as context-based mathematics courses taught in other departments; and as courses in other disciplines that employ significant (albeit often hidden) mathematical methods. I have no data to quantify the "biomass" of mathematics taught through these three means, but to a first approximation I would conjecture that they are approximately equal.

If that is true, then I can with some confidence suggest that the biomass of mathematics that is learned, remembered, and still useful five years later is decidedly tilted in favor of the hidden curriculum. Most of what is taught in the traditional mathematics curriculum is forgotten by most students shortly after they finish their mathematical studies, but most of what students learn in the context of use is remembered much longer, especially if students practice in the field for which they prepared.

Thus this paradox: As widespread use makes post-secondary mathematics increasingly essential for all students, mathematics departments find themselves playing a diminishing role in the mathematical education of postsecondary students. The obvious corollary poses another paradox: to exercise their responsibility to mathematics, mathematicians must think not primarily about their own department but about the whole university. Mathematics pervades not only life and work, but also all parts of higher education.

Rather than viewing Steen's three models as competing with one another, they can be seen as antecedents for new connections and conversations about curriculum. In "Eight Approaches to Teaching Mathematics," Jim Neyland (1995) offers a different taxonomy of mathematics education based on different pedagogical philosophies and various approaches to learning mathematical concepts. Neyland's paper provides scientists who are unfamiliar with educational research a convenient access point for constructivist principles that have been adopted by most of the major curricular reform initiatives of the past two decades. These initiatives offer one place to find commonalities and differences in recent conversations about teaching across these disciplines.

Some disciplines such as chemistry benefit from having one primary professional society—the American Chemical Society (ACS)—that can put enormous resources behind new initiatives. Other disciplines, such as mathematics, have a small number of major professional societies that can promote educational discussions: the Mathematical Association of America (MAA), the American Mathematical Society (AMS), and the Society for Industrial and Applied Mathematics (SIAM). Unfortunately, although biology has literally hundreds of major professional societies, there is no single society amongst them that could either speak for or listen to the majority of teaching biologists.

It may not be surprising, therefore, that the National Science Foundation funded systemic curriculum initiatives only in chemistry and mathematics. From these two initiatives and their subsequent follow-up activities, voluminous material has been produced that has enormous potential for being adapted in the post-*Bio 2010* era. By looking at early efforts in a variety of disciplines, biologists can benefit from the rich collection of curricular materials that have already been developed, many of which use mathematics and computers in biology. Second, and as important, those who undertake changes to reflect *Bio 2010* can learn from pedagogical experiments and associated educational research on problem solving that have taken place in different disciplines. A quick survey of projects in different scientific fields, including those illustrated with icons in Figure 4, follows.[11]

Six projects are recognized in this figure because of their potential. First, one of the oldest and finest collections of mathematical educational materials that uses biological examples for teaching has been the UMAP Journal published by the Consortium on Mathematics and its Applications (COMAP).[12] Hundreds of articles published in the Journal are also duplicated as separate modules for classroom use. Some of the subjects treated include:

> aquaculture of mussels
> chaotic models of blood cell populations
> combinatorics of DNA
> compartment models of cellular metabolism
> differential equation models of whales and krill
> epidemics
> exponential growth and decay
> forensic DNA tests for paternity or crime
> fractals
> genetic counseling
> Hodgkin-Huxley equations for nerve firing
> honeycomb design
> hypothesis testing in field biology
> information theory measures of biological diversity
> linear algebra for buffalo herd management
> Lotka-Volterra predator-prey model
> measuring cardiac output
> mortality functions
> phyllotaxis
> population genetics
> population projection

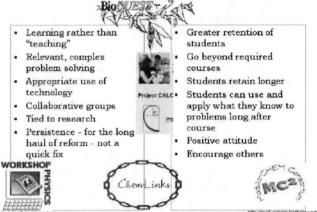

Figure 4. Common elements of reform among five STEM disciplines: biology, chemistry, mathematics, computational science, and physics.

> prescribing safe and effective drug dosages
> Ricker salmon model
> shoot development in plants
> Simpson's paradox on data analysis of populations
> visual perception

Modules can be purchased individually or in thematic collections such as a "Medipack," "Enviropack," or two different "Statpacks." In parallel with the undergraduate collection, COMAP also has other materials aimed at K-12 educators. I have found these to be a rich source of curricular materials for biology classrooms at the college level as well. These materials have been vetted in many classrooms and cover multiple areas within biology. They lay out the relevant mathematics clearly, and provide references to help access a broader literature.

Second, the major national systemic reform group in biology for the past 18 years has been the BioQUEST Curriculum Consortium.[13] In our various publications (*BioQUEST Library, Microbes Count!, BioQUEST Notes*) and extensive investigative case-based learning modules of our *LifeLines OnLine* project,[14] we have distributed a huge variety of field-tested simulations, tools, databases, and modeling environments across the breadth of undergraduate biology education. These materials include curricular resources for biochemistry, bioinformatics, cell biology, demography, developmental biology, ecology, epidemiology, evolution, genetics, molecular biology, neurobiology, physiology, and systematics. BioQUEST simulations are called "strategic

[11] mc2.cchem.berkeley.edu

[12] www.comap.com. See Appendix 3 for a list of COMAP modules.

[13] www.bioquest.org

[14] www.bioquest.org/lifelinesonline

simulations" because they are intended to help students learn long-term strategies of research. For a description of some of the mathematics represented in these software packages, see Jungck (1997a; 1997b).

The third program featured in Figure 4 is Project CALC (Calculus: A Laboratory Course), a pioneer in teaching calculus in terms of its application to everyday life.[15] Project CALC is especially rich in biological, environmental, medical, and public health examples. Their work was so foundational and transformative that their approach has been reinterpreted, modified, and toned-down by many other calculus education software and textbook developers, and adoptions in different colleges and universities have used such diverse symbolic manipulation packages as *Mathematica*, *Maple*, and *Derive*.

After its original funding ended, Project CALC formed the Connected Curriculum Project (CCP),[16] a consortial effort to archive on the web a variety of modular interactive learning materials for mathematics and its applications. CCP materials combine the flexibility and connectivity of the web with the power of computer algebra systems (CAS). (Materials are archived in various formats for different CAS systems.) A separate link points to a section of biological materials as well as to a Post-CALC project[17] which has produced interactive modules for high school students who have finished calculus. A related project at a more advanced mathematical level is the Consortium on Differential Equations Education (C*ODE*E).[18] Although this project is not longer active, many of their cases have substantial biological content and their archive is searchable on the web.

Third, the National Computational Science Institute (NCSI),[19] is a project of the Shodor Foundation in North Carolina that offers workshops and online educational materials for computational science, numerical modeling, and visualization. In contrast to BioQUEST's 3P's of "Problem Posing, Problem Solving, and Persuading Peers," the NCSI motto is "Pull, Push, and Permeate." Philosophically, their director, Robert Panoff, stresses the difference between simply using a computer to solve a problem or to explore a simulation and actually doing computing by writing equations, setting up spreadsheets,

and modeling dynamics with packages such as Berkeley Madonna or Stella or Extend. A related resource is the Special Interest Group on Computer Science Education (SIGCSE)[20] of the Association for Computing Machinery (ACM). SIGCSE offers annual technical symposia on computer science education; members can easily search an extensive set of archived reports from the proceedings of these meetings as well as numerous other publications devoted to education.

One of my favorite resources, now nearly thirty years old, is a bibliography of approximately two hundred references in computer science education (Austing, et al., 1977). The bibliography itself is preceded by brief descriptions organizing the references into categories: survey reports, activities of professional organizations, philosophy of programs, description of programs, description of courses, and other materials. From this era, I recall one of my favorite quotes: "Deep technology is of little value without a deep view of education" (Dwyer 1974). Dwyer's advice still seems highly relevant:

> Recent advances in technology, especially those related to the ideas of computer science and human problem solving, offer fascinating potential as agents for implementing a rich and quite deep view of education. Yet many educational programs involved with computer technology studiously avoid entanglement in such issues, preferring instead to accept the safe but shallow waters of drill and practice, frame-oriented tutoring, computer-aided testing, and other traditional management-of-student applications. [Thus] there has undoubtedly been a tendency to associate technology with some of the more mechanistic aspects of instruction.

All Dwyer would need to update such a reactionary list is to include dazzling students with 'edutainment' or monitoring them with clickers throughout an interactive large classroom lecture where the professor still is the disseminator of knowledge and the evaluator of answers.

A recent development in computer science education has been the creation of minors programs in computational science. These programs are very inviting to non-computer science majors who need to learn to use computers extensively in an interdisciplinary environment. Carleton University in Ottawa, Canada defines their version of this new approach as:

> The Computational Sciences are rapidly emerging new disciplines that connect the power of computer science with the leading edge of experimental work in the natural sciences. Using massive computing power, sophisticated algorithms and rapid access to vast databases, sci-

[15] www.math.duke.edu/education/proj_calc/
[16] www.math.duke.edu/education/ccp/
[17] www.math.duke.edu/education/postcalc/
[18] www.math.hmc.edu/codee/main.html
[19] www.computationalscience.org
[20] www.sigcse.org

entists can tackle important questions not approachable in any other way.

Computational Science brings computer science right down to the laboratory bench. A computational scientist is equally prepared to develop a new algorithm or perform experiments in the laboratory. With a solid background in computer science and the techniques and tools available to analyze experimental data, a computational scientist is also a fully qualified graduate in Biochemistry, Biology, Chemistry or Earth Sciences.[21]

This approach has been widely adopted at liberal arts colleges because it meets the needs of many students seeking a broad general education. The "Computational Science Across the Curriculum" consortium project headquartered at Capital University and lead by Thomas Gearhart and Ignatios Vakalis has been the U.S. national leader in this effort.[22] The opposite side of the coin—bioinformatics for computer sciences majors—is also a growing option at many institutions (Doom, et al., 1999).

The fourth highlight of Figure 4 is ChemLinks and the Modular Chemistry Consortium. When NSF developed a systemic initiative in undergraduate chemistry they supported five separate groups. Since the initial funding ended, these groups have collaborated in offering follow-on workshops to share the best of what they have learned. Two of these five, ChemLinks and the Modular Chemistry Consortium (MC2), combined to found Chemical Connections,[23] whose materials focus on the use of real-world problems like global warming and ozone holes. Modules from this project have been published in a variety of formats—print, CD, disk, and downloadable pdf form.[24] The aim of these modules is to provide inquiry-based classroom, laboratory, and media activities that help students develop the knowledge to answer real-world questions through student-centered, collaborative classroom activities.

Finally, there is Workshop Physics, begun by Priscilla Laws at Dickinson College in 1986.[25] This project has had profound impact on reforms in many other disciplines because of three critical components. First, lecture-free workshops consisting of three two-hour laboratory blocks per week replaced the traditional pattern of three or four lecture hours and three lab hours per week. Second, miniature probes for measuring force, motion, pH, temperature, etc. were developed and linked to soft-ware that made transparent connections to mathematics (e.g., by enabling students to visualize derivatives as tangents and integration as areas under curves). This instrumentation and software are easily employed in many scientific disciplines. Third, whenever possible the student became the object in experiments on force and motion so that they understood concepts like gyroscopic forces kinesthetically. At a time when many physics departments were almost entirely male and losing majors, adopters of Workshop Physics were able to recruit and retain more majors, many of whom were women. Dissemination of this project was greatly enhanced by a very cooperative community of reformers and education researchers in the physics community.

Complexities

Physicist Pierre-Gilles De Gennes, a 1991 Nobel laureate, and his popularizer Jacques Badoz (1996) warn us about the excesses of too much attention to mathematics in science education to the exclusion of other crucial components:

> *The Imperialism of Mathematics.* The Entrance Examination Theorem. There is a theorem that states: 'Whenever an entrance examination is instituted in a scientific discipline, it invariably becomes an exercise in mathematics.' … The student … is at risk of remaining disconnected from the real world during his entire life. … They may have learned to master certain tools, …, but they will suffer from crippling weaknesses in … observation, manual skills, common sense, and sociability.… [M]athematics … is a superb discipline, of which I am personally very fond.… What I object to is using math as a selection tool.… This discipline must, of course, be included in preparatory programs, but not to the point of becoming their primary ingredient.

While this commentary is particularly relevant to practices in France, it portends problems if we swing our pendulum to the opposite pole from currently having almost no mathematics and computer science in beginning biology. Another caveat recently has been raised by Tom Popkewitz in a lead article in a recent issue of the *American Educational Research Journal* (2004): He argues that (a) "the emphasis on 'problem-solving,' collaboration, and 'communities of learning' sanctify science and scientists as possessing authoritative knowledge over increasing realms of human phenomena, thus narrowing the boundaries of possible action and critical thought; " and (b) "while reforms stress the need for educational equity for 'all children,' with 'no child left behind,' the pedagogical models divide, demarcate, and exclude particular children from participation" (p.3).

---

[21] www.carleton.ca/natsci/compsci/whatis.html

[22] www.capital.edu/prosp/ug/computational-science.html

[23] chemlinks.beloit.edu

[24] wwnorton.com/college/chemistry/chemconnections/

[25] physics.dickinson.edu/~wp_web/WP_homepage.html

It seems likely that we will soon go through a period featuring enormous diversity of new programs linking biology, mathematics, and computer education. Participants in this curricular transition will represent different degrees of commitment to this shared responsibility—visitors, collaborators, team teachers, course linkage partners, serial tag teams, polymaths, gadflies, imperialists, etc.—and their behavior will vary accordingly. Will we rely on careful education research to distinguish among their approaches or will decisions be made primarily on ideological, political, or economic considerations?

So far no commercial publisher seems to fully understand how to support the use and sustained development of rigorous, classroom-tested, research-driven courseware in STEM disciplines. How will the transition to more mathematically and computationally intense biology education fare if publishers continue their current worry that "for every equation we lose 10% of our market"?

As worrisome, many departments have dropped laboratory expectations for courses due to increased costs and liability concerns. Yet contemporary pedagogical reforms all encourage much more activity-driven, collaborative exploration with professional tools applied to complex problems. Will these new programs even be allowed to germinate if administrators believe that the associated costs are too high?

A popular paradigm of the stages of curriculum reform has been called "Crossing the Chasm" (Figure 5)[26] after the Silicon Valley phenomenon of waves of technological appropriation by different constituencies (Rogers 1995, Moore 1991). The cautions cited above should not diminish the current zeal to finally bring mathematics and computer science into the biology curriculum beyond the experiments of a few isolated individuals whom this paradigm would characterize as innovators and early adopters.

Let us hope that we can "cross the chasm" and engage an "early majority" in significant reform of undergraduate STEM education. The NRC *Bio 2010* report has catalyzed interest. What is needed now, in addition to time and money, is the sustained will required to develop more quantitative, computationally rich problem solving and collaborative investigative curricula that genuinely help students learn and prepare them for contemporary professions.

What we have reported here is that: (1) there already exists the collective will of the educational arms of many

[26] www.spin.org/Myers/sld015.html

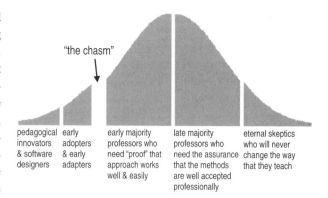

Figure 5. "The Adopter Continuum"

professional biological research societies and many of their leaders: (2) significant developments in interdisciplinary science education that embrace investigative, quantitative, collaborative, problem-solving-based learning have been made by curricular reforms initiatives over the past two decades and they form a community of peer support for others who want to engage in such work: (3) a large number of simulations, tools, and databases have been constructed to help students explore, model, and analyze complex scientific problems: and (4) there is a science and mathematics education literature as well as recent research in cognitive science that support the learner-centered and activities-centered curricula previously discussed. Thus, propitiously, the confluence of recommendations, research, reform initiatives, technologies, and available curricular materials should provide an excellent basis for moving ahead in addressing these challenges.

References and Related Reading

Adler, Fred R., *Modeling the Dynamics of Life: Calculus and Probability for Life Scientists*, Brooks/Cole, Pacific Grove, CA, 1998.

Austing, Richard H., Bruce H. Barnes, and Gerald L. Engel, "A survey of the literature in computer science education since curriculum '68," *Communications of the ACM* 20 (1977), no. 1.

Committee on the Undergraduate Program in Mathematics, *CUPM Curriculum Guide 2004: Undergraduate Programs and Courses in the Mathematical Sciences*, Mathematical Association of America, Washington, DC, 2004.
 www.maa.org/cupm

Bloom, David E. and Joel E. Cohen, "Education for all: an unfinished revolution," *Daedalus* 131 (Summer 2002), no. 3, 84–95.

Cornu, Bernard and Anthony Ralston, (eds.), *The influence of computers and informatics on mathematics and its teaching* (Science and Technology Education Series, 44), UNESCO, Paris, 1992. (ERIC Document Reproduction Service No. ED 359 073).

De Gennes, Pierre-Gilles and Jacques Badoz. *Fragile Objects: Soft Matter, Hard Science and the Thrill of Discovery*, Copernicus (Springer-Verlag), New York, NY, 1996, pp. 158–161.

Deckelman, Steve, "Using interdisciplinary curricula in a project-based investigative seminar for senior undergraduates: A case study in mathematical biology." *Council on Undergraduate Research Quarterly* (September 2003), 35–39.

Doom, Travis, Michael Raymer, Dan Krane, and Oscar Garcia, "A proposed undergraduate bioinformatics curriculum for computer scientists." *ACM SIGCSE Technical Symposium on Computer Science Education* (33rd: 2002). ACM Press, New York, NY, 2002, pp. 78–81.

Dwyer, Tom, "Heuristic strategies for using computers to enrich education." *International Journal of Man-Machine Studies* 6 (1974), 137–154.

Ganter, Susan and William Barker, editors, *Curriculum Foundations Project: Voices of the Partner Disciplines*. Mathematical Association of America, Washington, DC, 2004. www.maa.org/cupm/crafty/

Gross, Louis J., "Education for a biocomplex future." *Science* 28 (2000), 807.

———, "Quantitative training for life-science students," *BioScience* 44 (1994), 59.

Jungck, John R, "Ten Equations that Changed Biology: Mathematics in Problem-Solving Biology Curricula," *Bioscene: Journal of College Biology Teaching* 23 (May 1997), no. 1: 11–36. URL: papa.indstate.edu/amcbt/volume_23/.

———, "Biological Aftermath: What Can We Learn from Contemporary Mathematics Reform?" *BioQUEST Notes* 7 (March 1997), no. 2, 1–5, 8–13.

Jungck, John R., Ethel D. Stanley, and Marion Field Fass, editors, *Microbes Count! Problem Posing, Problem Solving, and Persuading Peers in Microbiology*. American Society for Microbiology Press, Washington, DC, 2003.

Jungck, John R., Stephen J. Everse, Patti Soderberg, Ethel D. Stanley, James Stewart, and Virginia Vaughan, (eds.), *The BioQUEST Library*, Academic Press, San Diego, CA, 2002.

Jungck, John R., Ethel Stanley, Sam Donovan, Patti Soderberg, and Virginia Vaughan, "'Crossing the Chasm' of Curricular Reform: BioQUEST Curriculum Consortium Builds Linkages to Diverse Communities," *UniServe CAL-laborate* 4, 2000. URL: science.uniserve.edu.au/pubs/callab/vol4/jungck.html.

Moore, Geoffrey A., *Crossing the Chasm,* Harper Business, New York, NY, 1991.

National Research Council, *Bio 2010: Transforming Undergraduate Education for Future Research Biologists*. National Academies Press, Washington, DC, 2003.

Neuhauser, Claudia, *Calculus for Biology and Medicine*. Upper Prentice Hall, Saddle River, NJ, 2000. Second Edition, 2004.

Neyland, Jim, "Eight Approaches to Teaching Mathematics." In Jim Neyland, (ed.), *Mathematics Education: A Handbook for Teachers, Volume 2*, National Council of Teachers of Mathematics, Reston, VA, 1995, pp. 34–48.

Popkewitz, Thomas, "The Alchemy of the Mathematics Curriculum: Inscriptions and Fabrication of the Child." *American Educational Research Journal* 41 (2004), no. 1, 3–34.

Ralston, Anthony, "Computer science, mathematics, and the undergraduate curricula in both," *American Mathematical Monthly* 88 (1981), 472–485.

Ralston, Anthony, "Research Mathematicians and Mathematics Education: A Critique." *Notices of the American Mathematical Society* 51(2004), no. 4, 403–411.

Rogers, Everett M., *Diffusion of Innovation*, (Fourth Edition), New York, Free Press, 1995.

Rothman, Brian, "Will the digital computer transform classical mathematics?" *Philosophical Transactions of the Royal Society: Mathematical, Physical and Engineering Sciences* 361 (2003), no. 1809, 1675–1690.

Stanley, E. and M.A. Waterman, "LifeLines Online," *Journal of College Science Teaching* (March/April, 2000), 306–310.

Steen, Lynn Arthur, "Revolution by Stealth: Redefining University Mathematics." In Derek Holton, Editor, T*eaching and Learning of Mathematics at the University Level*. Proceedings of the ICMI Study Conference, Singapore, 1998, Kluwer Academic Publishers, Dordrecht, 2000.

Thompson, D'Arcy Wentworth, *On Growth and Form*, Cambridge University Press, Cambridge, UK, 1917.

Tucker, Alan C., (ed.), *Models that Work: Case Studies in Effective Undergraduate Mathematics Programs,* Notes 38, The Mathematical Association of America, Washington, DC, 1995.

Waterman, M.A., "Investigative case study approach for biology learning," *Bioscene: Journal of College Biology Teaching* 24 (1998), no. 1, 3–10.

The "Gift" of Mathematics in the Era of Biology

Lynn Arthur Steen
St. Olaf College

For most of the twentieth century, mathematics was seen as a close and natural partner of physics and engineering—a "wonderful gift," in physicist Eugene Wigner's memorable words, that we "neither understand nor deserve."[1] The mainstream of secondary and postsecondary mathematics education was channeled to nourish the roots of the physical sciences. The general public, notably including students and their parents, believed that only those who wanted to be engineers really needed advanced mathematics. From trigonometry through calculus and on into advanced calculus and differential equations, three to four years of students' mathematical study in grades 10–14 was designed to support the parallel curriculum in physics. Indeed, until well into the second half of the twentieth century, most universities required all mathematics majors to take a year of calculus-based physics.

Now, however, midway through the first decade of the twenty-first century, science has moved on. Biology has replaced physics as the next "big thing"—the crucible of innovation—not only in science but also in mathematics. "The mathematics involved in studying the genome and the folding of proteins is deep, elegant, and beautiful—all words that often were reserved only for pure mathematics in the past century," notes the Executive Director of the American Mathematical Society. "The sophisticated blending of mathematics and biology already is a spectacular new area of research that is certain to grow enormously."[2]

Some reverse the image and argue that mathematics is the next big thing in biology. Biologist Eric Lander, a principal leader of the Human Genome Project, speaks of "biology as information," as a vast library filled with the "laboratory notebooks of evolution," one for every species, with chapters for every tissue, each containing genome sequences as well as expressions of RNA and proteins.[3] All these volumes are written in code that can be deciphered only by use of sophisticated mathematical algorithms. Once read, sequences can be mapped to functions, and thence to malfunction (disease) and to the evolution of function and form in both organisms and species. According to Lander, the genetic revolution in

Lynn Arthur Steen is Professor of Mathematics and Special Assistant to the Provost at St. Olaf College, Northfield, MN. A former president of the Mathematical Association of America, Steen is author of *Everybody Counts* (1989) and *Achieving Quantitative Literacy* (2004), and editor of *Calculus for a New Century* (1987), *Why Numbers Count* (1997), and *Mathematics and Democracy* (2001). Steen received his PhD in mathematics from MIT in 1965.

[1] Eugene Wigner, "The unreasonable effectiveness of mathematics," *Communications in Pure and Applied Mathematics*, 13 (1960) 1–4.

[2] John Ewing. "The Next Big Thing in Mathematics." *The Chronicle of Higher Education*, September 20, 2002.

[3] Eric Lander, "Beyond the Human Genome Project: Biology as Information." *Genes, Genomes, and Medicine*, Columbia University Digital Knowledge Ventures. URL: c250.Columbia.edu/genomes.

Supporting Research in Mathematical Biology

Revolutionary opportunities have emerged for mathematically driven advances in biological research. Examples include:

- Evolutionary theory and practice arising from genomics advances;
- Statistical approaches to the discovery of genes that contribute to complex behavior;
- Modeling of complex ecological systems;
- Explanatory and predictive models of the cellular state;
- Cellular growth, motility, division, and membrane trafficking;
- Metabolic circuitry and dynamics;
- Population dynamics;
- Signal transduction;
- Informational molecule dynamics;
- New algorithms for phylogenetic analysis;
- Dynamics of pattern formation in development and differentiation;
- New approaches to the prediction of molecular structure;
- Improved algorithms for x-ray crystallography, NMR and electron microscopy;
- Simulations of human systemic responses to burn, trauma and other injury;
- Systemic effects of pharmacological agents and their genetic & environmental modifiers.

The National Science Foundation (NSF) and the National Institutes of Health (NIH) collaborate to provide support for research in cross-disciplinary areas such as these. Proposals are expected to identify innovative mathematics or statistics needed to solve an important biological problem. Work supported under this initiative must impact biology and advance mathematics or statistics. Thus, collaborations between mathematical and biological scientists are expected.

Officially called the Initiative to Support Research in the Area of Mathematical Biology, this program is a joint activity of the Division of Mathematical Sciences (DMS) in the Directorate for Mathematical and Physical Sciences (MPS) and the Directorate for Biological Sciences (BIO) at the National Science Foundation (NSF) and the National Institute of General Medical Sciences (NIGMS) at the National Institutes of Health (NIH). Further information is available at: www.nsf.gov/pubsys/ods/getpub.cfm?nsf04572.

biology contains keys to "the most remarkable library of information on this planet."

The relatively sudden emergence of biology as the dominant scientific partner for mathematics in both research and education has created major challenges for both disciplines. Biology research—and with it the emerging multibillion dollar biotech industry—is hampered by lack of scientists able to work in teams where both biological and mathematical skills are required. Biology education is burdened by habits from a past where biology was seen as a safe harbor for math-averse science students. Biology faculty need to learn about the new quantitative tools while at the same time teach students who are often refugees from mathematics. And while embracing the multitude of research opportunities offered by the new biology, mathematical scientists who were educated in the physics paradigm face the daunting prospect of learning to teach new cross-disciplinary courses and to conduct collaborative research with colleagues in life science fields awash in unfamiliar methodologies, vocabulary, and theoretical foundations.

(Sidebars distributed throughout this paper illustrate a variety of responses to these challenges.)

The challenges for undergraduate education of what has come to be called the "New Biology" were forcefully articulated by the National Research Council in the widely acclaimed report *Bio 2010*.[4] This report unhesitatingly describes the digital transformation of biology as a "revolution," and offers far-reaching recommendations for how to synchronize undergraduate programs with the transformed reality of the new biology. The thesis of *Bio 2010* is that biology is not a separate science but an "integrative discipline" in which many aspects of the mathematical and physical sciences "converge to address biological issues."

Consequently, *Bio 2010* urges colleges and universities to create strong interdisciplinary curricula that integrate

---

[4] Committee on Undergraduate Biology Education, *Bio 2010: Transforming Undergraduate Education for Future Research Biologists*, National Research Council, National Academies Press, Washington, DC, 2003.

Workshops on the New Biology

A two-week intensive program introducing modern concepts in biology for mathematical and physical scientists took place in June, 2003 at the DIMACS Center, Rutgers University. Its goal was to introduce topics in molecular and cell biology that are relevant to those who wish to work at the interface of biology, mathematics, computer science, chemistry, and physics.

The first week, "The DNA Revolution," introduced the fundamentals of modern molecular biology, genetics, and biotechnology to participants who had little background in biology or biochemistry. It included a web-based primer of important biological information that covers chemical structures, biochemical processes, and classical genetics; a virtual laboratory with realistic exercises that illustrate the activities of modern-day bench scientists; and specially designed "hands-on" bioinformatics research projects. In addition, the tutorial examined scientific questions that interest present-day biomedical researchers, especially problems that can be addressed through quantitative approaches.

The second week reviewed cell biology—including energy generation, cell division, cell cycle control, cell communication, cytoskeleton, membranes, and intracellular compartments—and then summarized open questions. Major topics were illustrated by in-depth analyses of significant recent biophysical studies (e.g., mathematical models of cell and viral kinetics that explain the emergence of drug-resistant viruses). Participants became familiar with important websites useful to researchers and heard presentations on current research in fields at the interface of the biological, mathematical, and physical sciences.

Organizers: William Sofer, Rutgers University, sofer@waksman.rutgers.edu
              Paul Ehrlich, Rutgers University, pehrlich@lutece.rutgers.edu

mathematical, information, physical, and life sciences. Defying the long-standing tradition of biology being taught as the least mathematical science, *Bio 2010* documents biologists' need for a strong foundation in the mathematical and information sciences. Indirectly, *Bio 2010* also implies an urgent need for a parallel change in the education of mathematical scientists: to be prepared either to teach or conduct research, today's students in the mathematical sciences need a strong foundation in the life sciences—just as their predecessors needed a strong foundation in physics.

These recommendations are not just parochial issues of concern to a small number of departments and students. One in six entering college students now expects to major in a biological, health, or life science. The number of graduates who major in the biological sciences is now greater than those in engineering or psychology. The quantitative and integrative challenges posed by the new biology herald changes coming to all undergraduate science in the first part of the twenty-first century. This is indeed a Big Thing.

Quantitative Biology: Origins

Even without understanding details, most people are well aware of the deep connections between the mathematical and physical sciences—from gravity to relativity, from quantum theory to string theory. Yet very few recognize similarly deep connections between the mathematical and life sciences. Even today when decoded genomes share headlines with speculations about dark matter, most peo-

ple do not connect genomics with mathematical tools the way they do the theories of physics or astronomy. This bias in public perception—which influences students and schools—is in part the result of the physical-science orientation of traditional mathematics curricula. From their own education, adults have learned to see mathematics as comprising equations such as those used in physics (e.g., $E = mc^2$; $v = d/t$). But they have not learned to see mathematics as necessary for, or even relevant to, biology.

In one sense, this view is not unreasonable. With few exceptions, the mathematical and quantitative tools used to study biological processes have been only modestly effective, especially as compared with similar applications in the physical sciences. The reason is simple: Living organisms are vastly more complicated than lifeless matter. Quite naturally, the simpler problems of physics were solved before—a century or two before—those of biology. But now, at the beginning of the twentieth first century, scientists finally have received the "gift" of quantitative tools required to model biological processes with the same understanding as they have earlier achieved for physical systems.

In another sense, however, the widespread perception of biology as an exceptional science that does not speak the language of mathematics is a myth. Mathematics is the science of patterns,[5] and biology overflows with pattern. Visible examples are most obvious: spiral patterns in nautilus shells; intricate geometry of radiolaria skele-

[5] Lynn Arthur Steen, "The Science of Patterns," *Science* 240 (29 April 1988), 611–616.

tons; hexagonal structure of honeycombs; patterns on animal skin and butterfly wings; bilateral symmetry of most animal species; and much more.[6]

The great Scottish zoologist D'Arcy Thompson was the first person to argue persuasively that such patterns are due to mathematics, not instinct or genetics. "There are no exceptions to the rule that God always geometrizes. The problems of form are in the first instance mathematical problems, the problems of growth are essentially physical problems." In his classic monograph On Growth and Form,[7] Thompson used established principles of mathematics and mechanics to analyze skeletons and structure of living organisms. Through these examples he documented how the trail of evolution reflected the necessity of physical law, thus providing an explanation for how different life forms took on their varied shapes. His larger contribution was to break down the unstated assumption of the time that living things and inanimate objects occupied different scientific realms and were subject to different laws. For Thompson, both obeyed the laws of mathematics.

Other patterns are less visible but no less fundamental. During the eighteenth and nineteenth centuries, quantitative characteristics became increasingly important as measures of social well being. Censuses became common as a means of counting people and wealth, noting births, deaths, and immigration. Businesses counted and weighed natural assets such as wheat, timber, and livestock. World trade came to depend on projections of the size of natural populations such as wild game, crop harvests, and human population. Thus not only scientists but also business and political leaders all had increasing interest in quantification, particularly in the less visible patterns created by numbers that tally populations.

Surely one of the most widely recognized applications of mathematical principles to biology is Thomas Malthus' warnings about the calamitous consequences of population growth. Quoting himself in 1798, Malthus wrote "I said that population, when unchecked, increased in a geometrical ratio, and subsistence for man in an arithmetical ratio."[8] Malthus' "geometrical ratio" we now call exponential growth; even at modest rates, such growth is not realistic in the long run for any natural population. However, it is a very good model in certain circumstances, and is still the normative metaphor for discussions of population growth.

Another widely known early foray into using quantitative methods to gain biological insight is Gregor Mendel's meticulous documentation in the 1860s of numerical patterns in inherited characteristics.

"The ratio of 3:1, in accordance with which the distribution of the dominant and recessive characters results in the first generation, resolves itself ... into the ratio of 2:1:1. ... Since the members of the first generation spring directly from the seed of the hybrids, it is now clear that the hybrids form seeds having one or other of the two differentiating characters, and of these one-half develop again the hybrid form, while the other half yield plants which remain constant and receive the dominant or the recessive characters in equal numbers."[9]

Mendel's work not only described inheritance quantitatively, but made it possible to infer the notion of a gene—a minimal unit of heredity—long before physical genes were discovered. (This power of mathematics to predict scientific discoveries is not uncommon: the planet Neptune was predicted and ultimately discovered based on minute discrepancies between the theoretical and actual orbit of Uranus.)

Subsequently, arguments erupted over how recessive characteristics could survive eons of evolution. In 1908 British mathematician G. H. Hardy used simple high school mathematics—the binomial expansion—to infer from Mendel's analysis that the proportion of genes in a stable population will remain constant from generation to generation. "In a word, there is not the slightest foundation for the idea that a dominant character should show a tendency to spread over a whole population, or that recessive should tend to die out."[10] Now called the Hardy-Weinberg Law, this purely mathematical analysis helped resolve one of the early scientific objections to Darwin's theory of evolution.

More than a century earlier, the great eighteenth century Swiss mathematician Leonard Euler had established a similar stability result for the age distribution of populations. Uneven distribution of ages leads to uneven rates of reproduction of the population as a whole and can create "baby booms" and subsequent "busts." Euler showed that in the absence of external events such as immigration, the overall rate of growth of a population will even-

[6] John A. Adam, *Mathematics in Nature: Modeling Patterns in the Natural World*, Princeton University Press, Princeton, NJ, 2003.

[7] D'Arcy Wentworth Thompson, *On Growth and Form*, Cambridge University Press, Cambridge, UK, 1917, 1942.

[8] Thomas Malthus, *An Essay on the Principle of Population*, J. Johnson, St. Paul's Churchyard, London, 1798.

[9] Gregor Mendel. "Versuche über Pflanzen-Hybriden." 1865. English Translation: "Experiments in Plant Hybridization." www.biologie.uni-hamburg.de/b-online/e08_mend/ mendel.htm.

[10] G. H. Hardy, "Mendelian Proportions in a Mixed Population," *Science*, 28:706 (Jul 10, 1908), 49–50.

Supporting Undergraduate Biology and Mathematics

The explosion of knowledge in the life sciences over the past twenty years cuts across all levels from molecules to ecosystems. Current research is often characterized by integrative and interdisciplinary approaches. At the center of this explosion of knowledge is a revolution in instrumentation, computational abilities, information systems, and mathematical tools.

A parallel growth in understanding has taken place in the mathematical sciences. Theoretical advances in complexity, dynamical systems, and uncertainty, coupled with advances in modeling and computational methods, have helped mathematicians and statisticians put ideas into action. These advances have expanded use of mathematics and statistics beyond the traditional fields of physical science and engineering. As that expansion has taken hold, the life sciences and other fields are posing new kinds of questions for the mathematical sciences, stimulating further the growth of mathematical ideas.

Thus the intersection of the biological and mathematical sciences is a fertile field for both sets of disciplines, where results in each area lead to advances in the other. However, there are comparatively few people able to work in this intersection. The Undergraduate Biology and Mathematics (UBM) program at the National Science Foundation (NSF) is designed to attract and prepare students for careers in this important crossroads of two major disciplines—mathematics and biology. UBM programs are expected to:

- Be grounded in research activities involving both mathematical and biological sciences;
- Connect to regular academic studies, influencing the direction of academic programs for a broad range of students.
- Involve students from both areas in significant research experiences that connect to research at the intersection of the disciplines; and
- Show commitment to joint mentorship by faculty in both fields.

Individually, UBM projects will have a significant impact on the undergraduate programs of participating institutions. Collectively, they will strengthen the nation's research enterprise by providing new mechanisms for attracting a larger, more diverse group of students to careers that involve both the mathematical and biological sciences. Within this context, there is room for a variety of possible emphases, ranging from undergraduate research participation, through curriculum and faculty development, as well as internships outside the academic institution.

The Undergraduate Biology and Mathematics program is a joint effort of the Education and Human Resources (EHR), Biological Sciences (BIO), and Mathematical and Physical Sciences (MPS) directorates at the National Science Foundation (NSF). Further information is available at: www.nsf.gov/pubsys/ods/getpub.cfm?nsf04546.

tually stabilize: left to themselves, population distortions will subside. Euler did this by creating and solving a simple system of equations that represents the reproduction and survival rates for each age cohort.

But as Malthus argued, stable population growth cannot survive in the long run. Early in the nineteenth century the Belgian scientist Pierre François Verhulst suggested that each population has a theoretical long-term maximum that he called its "carrying capacity." As the population approaches its maximum, competition for resources tends to limit growth, thereby introducing negative feedback between size of a population and its rate of growth. In Verhulst's model,[11] the graph of population growth is no longer the dramatic (but unrealistic) unbounded exponential curve, but the "logistics" graph

that begins exponentially but then tapers off as it approaches the carrying capacity.

As "survival of the fittest" theories came under scrutiny in the period following publication of Darwin's *The Origin of Species,* mathematically-minded biologists extended Euler's and Verhulst's equations to multiple species competing in a single ecosystem. In this endeavor, scientists consciously imitated the models of physical systems that had proven so successful in engineering and mechanics. For example, Alfred Lotka at Johns Hopkins University employed simultaneous difference and differential equations to represent the behavior of a complex ecosystem as a set of trajectories in *n*-dimensional space. Lotka studied evolution as "the mechanics of systems undergoing irreversible changes." Indeed, the sections of his 1924 monograph on what he called "physical biology" borrowed thematic headings from mainstream mechanics textbooks of that era: statics, kinetics, dynamics.[12]

---

[11] Pierre François Verhulst. "Recherches mathématiques sur la loi d'accroissement de la population." *Nouv. mém. de l'Academie Royale des Sci. et Belles-Lettres de Bruxelles,* 18 (1845) 1–41. "Deuxième mémoire sur la loi d'accroissement de la population." *Mém. de l'Academie Royale des Sci., des Lettres et des Beaux-Arts de Belgique,* 20 (1847) 1–32.

---

[12] Alfred J. Lotka, *Elements of Physical Biology,* 1924; Reprinted as *Elements of Mathematical Biology,* Dover, 1956.

Verhulst's Model for Population Growth

In a capacity-limited population, the intrinsic rate of growth is diminished by the proportion of the capacity that has been used up. In symbols, as the population N approaches its maximum K, the rate of growth r will be reduced by $(N/K)r$. Thus the equation of unrestrained exponential growth

$$N(t + 1) = N(t)e^r$$

is replaced by

$$N(t + 1) = N(t)er(1 - N(t)/K).$$

Equivalently, the Verhulst model is often represented by the differential equation

$$dN/dt = rN(K - N)/K = rN(1 - N/K).$$

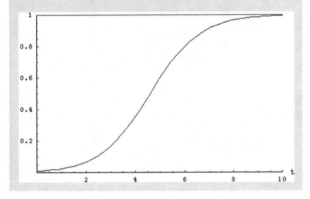

The most common application of Lotka's multi-species equations is to predator-prey situations where a dominant predator can so pressure its prey that the predator runs out of food, thence the predator population declines. With fewer predators, the prey rebound, and the cycle begins again. Classic long-term data documenting this phenomenon came from records of populations of Canadian lynx and snowshoe hares maintained for a century or more by the Hudson Bay Company. A similar analysis was carried out about the same time by the famous Italian mathematician Vito Volterra and his biologist son-in-law who was studying the stock of fish species in the Adriatic. Consequently, the predator-prey equations are now called the Volterra-Lotka equations.

Although these dynamical system models do a reasonable job of imitating certain patterns seen in actual populations (e.g., cycles of boom and bust), they are rarely very good at predicting events in real ecosystems. In one important study using single cell organisms competing for food and space, the Russian microbiologist G. F. Gause elaborated important differences between mathematicians' view of the Darwinian "stuggle for existence" and those found in nature. Gause showed, for example, that cyclic behavior is not an inherent charac-

teristic of predator-prey competitions, but often arises in natural situations as the result of some external factor.[13]

Gaps between theoretical models and experimental reality remain high in the life sciences. Assessing the size of populations is an inherently mathematical task—far more promising as a quantitative activity than, for instance, assessing the health of individuals or the function of genes. Mathematicians, statisticians, and biologists have worked for a century to develop effective mathematical models in biology, and population models in ecology have proved to be one of the most productive areas. Yet the famed British biologist J. Maynard Smith notes that even here, in the most promising of subfields, "it is usually better to rely on the judgment of an experienced practitioner than on the predictions of a theorist."[14]

Quantitative Biology: Expansion

These classic examples from the nineteenth and early twentieth centuries demonstrate that the image of biology as a non-quantitative science is a bit misguided. Contrary to popular myth, biology *is* amenable to quantitative methods—some areas more so (e.g., epidemiology, demography), others less so (cell biology, physiology). Moreover, just as Newton's effort to understand gravity led him to the development of calculus, scientists who investigate biological problems often find themselves inventing new mathematics (e.g., the Volterra-Lotka equations).

One indication of the breadth of intersection between mathematics and biology can be gleaned from the subject headings of *Mathematical Reviews* under the topic of biology. These span the entire spectrum of biology from molecular to ecological:

Animal behavior	Genetics
Biochemistry	Medical applications
Biomechanics	Medical epidemiology
Biomedical imaging	Molecular biology
Biophysics	Molecular structure
Biostatistics	Neural biology
Cell biology	Neural networks
Cell movement	Pharmacokinetics
Cellular processes	Physiological flow
Developmental biology	Physiology
DNA sequences	Plant biology
Ecology	Population dynamics
Enzyme kinetics	Protein sequences
Epidemiology	Signal processing
Evolution	Taxonomy

[13] G. F. Gause, *The Struggle for Existence*, The William & Wilkins Co., 1934; Dover, 1971.

[14] J. Maynard Smith, *Models in Ecology*, Cambridge University Press, 1974.

Environmental Biocomplexity

Complexity pervades biology and puzzles mathematicians. From matching segments of DNA to tracing electrical waves in the heart, the problems posed by biology often move well beyond the methods mathematicians have yet invented. Environmental science runs into the brick wall of complexity before it has rounded the first turn.

From the scientific perspective, environmental problems require integration of ever-changing data from biological, physical, and social systems. For example, ameliorating the impact of an airborne pollutant requires knowledge of its impact on the body, of its dispersal in the atmosphere, of the human causes of the pollution, of its economic and social costs, and of the consequences of any proposed program to bring about a change. Any one of these investigations alone would be complicated enough; together they merit the special label of "environmental biocomplexity."

From a mathematical perspective, complexity means something more than merely complicated. Phenomena that exhibit nonlinear dynamic behavior, or random fluctuations, or interactions of different scale are prone to chaotic behavior which is characterized by a special kind of unpredictability in which minute differences beyond our ability to observe or measure create noticeable differences in future phenomena that may be rather distant in time or space. This kind of complexity imposes limits on our ability to understand causes and to predict consequences.

The hallmark of a vital link between science and mathematics is a symbiotic relationship in which each draws strength from the other—posing challenges, raising questions, suggesting promising avenues of investigation. Many examples can be cited to illustrate that biological problems have stimulated important advances in the mathematical sciences, even as mathematical models have slowly improved in their ability to address biological problems with success and insight:

- In 1874, Francis Galton and H.W. Watson reported public concern about "the decay of families of men who occupied conspicuous positions in past times ... Surnames that were once common have since become scarce or have wholly disappeared." To test the widespread opinion that a "rise in physical comfort and intellectual capacity is necessarily accompanied by diminution in fertility," Galton and Watson developed a stochastic branching process to determine the likelihood of extinction of surnames under purely random conditions. They determined that this common conclusion had been "hastily drawn,"[15] and in the process introduced a new quantitative tool for use in analyzing evolutionary behavior.

- Later, in trying to understand how strongly the characteristics of one generation were evident in succeeding generations, Galton developed rudimentary ideas of regression and correlation.[16] These ideas were placed on firm mathematical footing by his younger colleague

Karl Pearson.[17] The most commonly used tools of today's statistics (correlation, regression) are direct descendents of ideas introduced in these experiments.

- Imaging technology has made great strides in the last two decades and has become an indispensable tool both for biology and medicine. Many factors combined to bring this about, not least advances in sensitivity of radiation detectors and increases in speed and power of computers that reconstruct three-dimensional images from two-dimensional data. Undergirding these modern computer methods is a strategy developed nearly half a century earlier by the Czech mathematician Johann Radon that showed, in essence, how to reconstruct images of objects from pictures of its shadows.[18] This Radon Transform is the central idea that made Computed Axial Tomography (CAT scans) possible.

- One of the more widely known examples of biological questions leading to new mathematics involves Benoit Mandelbrot's 1975 invention of fractals, abstract objects of fractional dimension that offer better models for life forms (trees, leaves, veins, lungs) and other natural shapes (coastlines, mountain ranges) than do the shapes of Euclidean geometry or curves of Newtonian calculus.[19] Iteration, a mathematical process that mimics the way natural organisms grow,

[15] Francis Galton and H. W. Watson, "On the probability of the extinction of families," *Journal of the Anthropological Institute*, (1874) 134–144.

[16] Francis Galton, *Natural Inheritance*, Macmillan and Co., London, 1889.

[17] Karl Pearson, "Mathematical Contributions to the Theory of Evolution. III. Regression, Heredity and Panmixia," *Philosophical Transactions of the Royal Society of London*, 187 (1896) 253-318.

[18] Johann Radon. "Über die Bestimmung von Funktionen durch ihre Integralwerte längs gewisser Mannigfaltigkeiten," *Berichte über die Verhandlungen der Sächsischen Akademie*, 69 (1917) 262-277.

[19] Benoit Mandelbrot, *Les objets fractals: forme, hazard et dimension*, Flammarion, Paris, 1975.

is the key to fractals. Notwithstanding that fractals have become part of popular culture and *New Yorker* cartoons, their utility has been thoroughly demonstrated in many parts of science.

- For decades, randomized double blind experiments have been the "gold standard" of clinical studies of proposed treatments for disease. Increasingly, however, patient advocates—often including the investigative scientists themselves—chafe under rigid protocols that deny some patients access to promising drugs until after the experiment is completed. An influential coterie of medical scientists are now using Bayesian statistics—a controversial eighteenth century approach that takes into account accumulating results[20]—to reconcile traditional statistical inference with current demands of ethics.[21]

- One of the many mysteries of biology that mathematics has helped explain is how a small homogeneous ball of embryonic cells can develop the dramatic differentiated patterns of a leopard's skin or a butterfly's wings. The answer, first formulated by the brilliant British computer theoretician Alan Turing, can be found in the interaction of chemicals whose concentrations during morphogenesis (embryonic development) alternately inhibit or activate each other. Turing used "reaction-diffusion equations" for these interacting chemicals to show how it is possible for patterns such as those found on skin to be created during the development of an embryo.[22]

- The mystery of morphogenesis extends from two-dimensions of skin patterns to differentiation of tissues and overall shape of the three-dimensional organism. What physical processes could conceivably enable such variety to emerge from uniform beginnings? Biologists can explain differences in tissue function by genes that are switched on or off, but this doesn't resolve the geometric puzzle about control of tissue boundaries. In searching for clues to this enigma, French mathematician René Thom, "one of the most original mathematical thinkers of the twentieth century,"[23] extended Turing's idea of biochemical attractors and derived seven possible geometric forms of morphogenesis that he termed "catastrophes."[24]

- Policy challenges associated with management of renewable resources has led to a whole new interdisciplinary field known as mathematical bioeconomics.[25] Fisheries, for example, need limits on allowable catches, seasons, and locations in order to ensure a stable population of fish as well as sufficient harvest to make the industry profitable. This has led to development of various models of optimal control theory that take into account variables such as aging, multiple species, capital investment, and stochastic events.

- Even though the heart is often viewed as a mechanical pump, reality is complicated by interaction of blood with flexible heart valves and muscular heart walls. Charles Peskin and colleagues at New York University have pioneered a mathematical model that regards the cardiac tissue as a part of the fluid in which additional elastic forces are applied. With the aid of an innovative "immersed boundary" method created to solve the equations of fluid motion under these circumstances, the NYU group can view a simulated beating human heart in order to study how it works, and how during heart attacks, it fails.[26]

- Shortly after defibrillators became visible in television emergency rooms, they began appearing on airplanes and are now being promoted for well-equipped home medicine chests. But no one can explain exactly why they work. James Keener at the University of Utah is one of several mathematicians attempting to build effective mathematical models of how electrical waves move across the heart to induce healthy rhythmic beats.[27] So far neither mathematicians nor biologists seem to have figured it out, but together they are making much more progress than either could alone.

This inventory of interaction between mathematical and biological sciences could go on for pages. However, those cited here are sufficient to document several important features of the relationship:

[20] Thomas Bayes, "Essay towards solving a problem in the doctrine of chances," *Philosophical Transactions of the Royal Society of London,* 53(1763) 370–418; reprinted in *Biometrika,* 45:3–4 (Dec. 1958) 293–315.

[21] Jennifer Couzin, "The New Math of Clinical Trials," *Science,* 303:5659 (6 Feb. 2004) 784–786.

[22] Alan Turing, "The Chemical Basis of Morphogenesis." *Philosophical Transactions of the Royal Society of London,* B237 (1952) 37–72.

[23] Jean-Pierre Bourguignon, "René Thom: 'Mathématicien et Apprenti Philosophe,'" *Bulletin of the American Mathematical Society,* 41:3 (July 2004) p. 273–274.

[24] Rene Thom, "Une théorie dynamique de la morphogénèse," In E. H. Waddington (Ed.), *Towards a Theoretical Biology,* Vol. I: Prolegomena, Edinburgh University Press, Edinburgh, 1968, pp. 152–166.

[25] Colin W. Clark, *Mathematical Bioeconomics: The Optimal Management of Renewable Resources,* John Wiley & Sons, New York, 2000.

[26] Frank C. Hoppensteadt and Charles S. Peskin, *Modeling and Simulation in Medicine and the Life Sciences,* Second Edition, Springer-Verlag, New York, 2002.

[27] Dana Mackenzie, "Making Sense of a Heart Gone Wild," *Science,* 303:5659 (6 Feb. 2004) 786–787.

From Molecular Regulatory Networks to Cell Physiology

John Tyson, Department of Biology, Virginia Tech

The living cell is a miniature biochemical machine that harvests material and energy from its environment and uses them for maintenance, growth and reproduction. These processes are carried out by macromolecular machines (enzymes, transport proteins, structural proteins, motor proteins, etc.) whose structures are encoded in nucleotide sequences (DNA and mRNA). The activities of these macromolecules are controlled and coordinated by regulatory networks of great complexity and exquisite effectiveness. These networks collect information from inside and outside the cell, process the data, and direct cellular responses that foster the survival and reproduction of the cell.[i]

How these regulatory systems work is no more or less apparent from their network diagrams than is a complex piece of electronics from its schematic wiring diagram. In the same way that electrical engineers create accurate mathematical representations of wiring diagrams and use these equations to design new devices, cell biologists now recognize the need for quantitative (mathematical and computational) modeling in order to understand how molecular regulatory systems control cellular responses. Only with this understanding can they re-engineer these control systems for industrial and medical purposes.[ii]

The regulation of growth, DNA synthesis, mitosis and cell division in eukaryotes is a prime example of what is called the "new biology" because of the central role of the cell cycle in all of cell physiology, because much is known about the underlying molecular regulatory system, and because of the immense importance of cell reproduction in health (cancer) and biotechnology (tissue engineering). A research group led by John Tyson (Virginia Tech) and Bela Novak (Budapest Univ. of Technology & Economics) uses nonlinear differential equations to represent wiring diagrams and provide reliable connections between the molecules and cell cycle physiology.[iii] By mathematical analysis (bifurcation theory) and numerical simulation, these researchers are building comprehensive, accurate, predictive models of cell cycle regulation in yeast cells, frog eggs, and multicellular organisms.

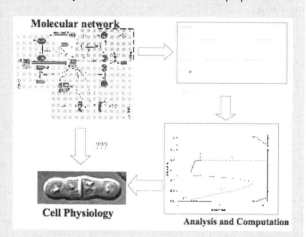

Research of this sort requires a sophisticated understanding of many fields: physiology, molecular cell biology, biochemical kinetics, dynamical systems theory, and numerical analysis, for starters. To provide such training, graduate programs in "computational biology" are springing up across North America, but the next generation of quantitative life scientists must start their preparation earlier. New interdisciplinary curricula for intelligent and highly motivated undergraduates are desperately needed. Such curricula could allow carefully chosen and guided students to put together courses from diverse fields, rather than to meet the expectations of traditional academic disciplines. To provide some structure, the program of study should be directed toward a specific research project in the final year. The exact combination of courses chosen by any student is not so important as is the interdisciplinary nature of the experience, which gives students confidence to cross boundaries in order to tackle cutting edge problems in the life sciences.

[i] D. Bray, "Protein molecules as computational elements in living cells," *Nature* 376 (1995) 307–312.

[ii] L.H. Hartwell et al., "From molecular to modular biology," *Nature* 402 (1999) C47–52.

[iii] J.J. Tyson et al., "Network dynamics and cell physiology," *Nature Rev. Mol. Cell Biol.* 2 (2001) 908–916.

1. Quantitative methods have always been part of biology.
2. Mathematical and statistical models are employed in biology at all scales from microscopic (molecular) to macroscopic (environmental).
3. All realms of the mathematical sciences—geometry, statistics, algebra, computer science, analysis, probability—share significant boundaries with the biological sciences.
4. Biological questions have motivated major advances in many parts of the mathematical sciences.
5. Finally, as biologists amass increasing amounts of data, the numerous connections between mathematical and biological sciences are becoming stronger and more essential.

In short, even apart from the genomics revolution, the gap between mathematical theory and biological reality is rapidly closing. After a century's struggle, mathematics has become the language of biology.

The New Biology: Data, not Disciplines

All the examples mentioned above share one further feature: they have nothing to do with genomics or bioinformatics. They are about all the other parts of biology. Indeed, no part of the biological terrain has been immune from the influence of statistics and mathematics: from molecules, cells and organs to organisms, populations and ecosystems, quantitative methods have been put to productive use in biology for well over a century, and continue so to this day.

Nevertheless, these models were rarely able to describe or predict with sufficient precision to answer specific research questions, nor were they able to demonstrate theorems of power comparable to those that undergird mathematical physics. Biological models were, quite literally, mostly of theoretical interest. More often than not, they suggested how things might behave rather than demonstrating how they would behave; they created hypotheses to test rather than conclusions to verify. For this reason, throughout the twentieth century mathematical biology was very much a niche discipline, fundamental to neither field and typically overlooked by both.

In recent years, of course, all this has changed. One reason is the recognition of genomes as the Rosetta Stone of life. For the same reason that the National Security Agency employs hundreds of mathematicians to crack intelligence codes, the biological research community now needs thousands of mathematical scientists to read information hidden in the genetic code.

A second reason is the powerful role of computer-enhanced visualization for research, diagnosis, and treatment. No longer are we limited to human eyesight enhanced by optics alone. Now we can "see" inside the body, even to the point of studying the shapes of molecules and witnessing internal organs. There are good evolutionary reasons why our minds can grasp more through a visual image than through words or data. Images enhance enormously each investigator's ability to "see," and therefore to understand.

A third reason is the revolutionary impact of networked databases. Data from millions of experiments reside in hundreds of thousands of computers all over the world. Like detectives who use networked fingerprint databases to solve decades-old crimes, biological investigators can now use the Internet to make connections and draw scientific conclusions from a vast international data warehouse. Networking magnifies enormously each investigator's scientific reach and resources.

The currency of the new biology is data—terabytes of data that are beginning to fill Eric Landers' imagined library of life. Traditional mathematical and statistical methods such as those described above—the cornerstones of twentieth century science—relied on a "reductionist" belief that the nature of complex processes can be explained by, derived from, or reduced to simpler and more fundamental laws. Examples of this paradigm include the triumphs of classical physics—Kepler's, Newton's, and Maxwell's laws—and the explanatory theories of multiple regression and factor analysis in subjects as wide ranging as evolution and education.

But in the new biology, data is king. In addition to verifying predictions of a mathematical model or statistical hypothesis (continuing the "old" biology), much current research is carried out by direct examination of data. Reversing the cycle of the "scientific method" in which theories are tested against data, these new methods begin (and often end) with data. The deluge of sequencing

Creating Courses that Integrate Biology and Mathematics

A one-week workshop in July, 2003 at Hope College in Holland, Michigan focused on courses and materials that integrate mathematics and biology in ways that benefit students from both disciplines. Participants consisted primarily of interdisciplinary teams including faculty from both the biological and mathematical sciences.

The workshop was designed to help participants better understand the uses of mathematics in current biological research and to help them develop appropriate interdisciplinary courses on their own campuses. It focused on key steps in this process:

- Establish collaborations between mathematicians and biologists;
- Create a course that attracts students from both disciplines;
- Find and develop curricular resources; and
- Incorporate appropriate software and/or experiments.

Participants received a variety of supporting materials, including resources used in existing mathematical biology courses, biology research papers that use quantitative methods, and the NRC report *Bio 2010*. Each participant completed a pre-workshop assignment from these resources. During the workshop participants developed a course syllabus and supporting curricular materials, worked through sample biology experiments, and completed sample computer exercises using modeling software such as STELLA. Participants are collectively creating a resource guide to be shared with others interested in developing similar courses.

Organizers:

Janet Andersen <andersen@hope.edu>, Mathematics, Hope College, Holland, MI
Eric Marland < marlandes@appstate.edu>, Biology, Appalachian State University, Boone, NC

information has already led to improvement in crops, treatment of diseases, and revision in our understanding of the tree of evolution. Genetic data are used to prosecute criminals and document the migration of peoples; digital pictures help researchers find drugs that bind to biologically active sites; and networked computers enable specialist consultation across cities and continents.

Data, however, do not honor disciplinary boundaries. They are what they are, without labels that say "biology" or "physics" or "mathematics." So a data-intensive science tends to be an interdisciplinary science. Indeed, many signs suggest that science disciplines in general are converging, "drawn together by common mathematical and computational paradigms" and because areas of greatest current interest "transcend traditional academic disciplines."[28] Unquestionably, these changes are having their strongest impact in biology:

> How biologists design, perform and analyze experiments is changing swiftly. Biological concepts and models are becoming more quantitative and biological research has become critically dependent on concepts and methods drawn from other scientific disciplines. The connections between the biological sciences and the

physical sciences, mathematics, and computer science are rapidly becoming deeper and more extensive.[29]

Thus the educational challenge of the new biology has two related but somewhat different dimensions. One is to recreate undergraduate biology as an interdisciplinary science, the other, to build high-capacity two-way bridges to the mathematical sciences.

Focusing on What Works

Changes in biology compel corresponding changes in biology education. Two fundamental issues are of central scientific importance:

- How to prepare students for a world where scientific boundaries are of diminishing importance.
- How to integrate the biological and mathematical sciences in students' undergraduate education.

A third is of utmost practical importance, namely:

- How to increase the number of biology students who are mathematically proficient and experienced in interdisciplinary work.

This latter issue is a direct consequence of the increasing importance and changing nature of biology. If biology

---

[28] Judith A. Ramaley. "Meeting the challenges in emerging areas." Keynote Lecture, Meeting the Challenges workshop, Feb. 27, 2003. URL: http://www.maa.org/mtc/.

[29] *Bio 2010,* op cit, p. 1.

What Works in Undergraduate Science Education

A thriving 'natural science' community is an environment where:

Learning is experiential and steeped in investigation from the very first courses for all students through capstone courses for students majoring in science, technology, engineering, and mathematics (STEM).

Learning is personally meaningful for students and faculty, making connections to other fields of inquiry, is embedded in the context of its own history and rationale, and suggests practical applications related to the experience of students.

Learning takes place in a community where faculty are committed equally to undergraduate teaching and to their own intellectual vitality, where faculty see students as partners in learning, where students collaborate with one another and gain confidence that they can succeed, and where institutions support such communities of learners.

— *What Works: Building Natural Science Communities*, Vol. I.
Project Kaleidoscope, Washington, DC, 1991.

were only an esoteric corner of science of limited practical importance, society could probably rely on the few who emerge from the current educational system with strong backgrounds in both mathematics and biology. But the opposite is true. It is not only the science of biology but also its manifold applications in agriculture, medicine, biotechnology and now bioterrorism that will suffer without a significant increase in the number of graduates who are professionally literate in both the mathematical and biological sciences.

Questions about better approaches to biology education are an important part of higher education's broader need to revitalize in the face of challenges posed by changing world conditions. One expression of this agenda is the Greater Expectations undertaking of the Association of American Colleges and Universities (AAC&U).[30] The greater expectations urged in this project are motivated by the increasing interconnection of knowledge in a rapidly changing world that knows no disciplinary boundaries. In many respects, they extend the expectations of the New Biology across the curriculum.

A similar but more focused effort has urged educators in grades 10-14 to pay greater attention to quantitative literacy (QL) for all students.[31] As a deluge of data has forced biology to embrace more sophisticated mathematical tools, so increased reliance on data in other fields is having similar effects. If more students finished secondary school with good quantitative tools, college biology courses could incorporate models requiring a greater degree of mathematical sophistication. Even fluency and comfort with ratios and percentages would be a positive step, not to mention more advanced topics. As important, if biology courses embraced mathematical and quantitative models of biological phenomena, they would make their own significant contribution to students' quantitative literacy. Two support networks seek to advance the QL agenda: the National Numeracy Network[32] and a Special Interest Group on Quantitative Literacy[33] of the Mathematical Association of America.

Experience teaches that we cannot predict very well who among the 1.5 million entering college students will have the motivation and capacity for productive careers in the life and health sciences. Many students who flee science and mathematics have as much potential as those who stay.[34] Current patterns of early filtering—first select for science interest, then separate mathophiles from mathophobes—hardly works for the old biology, and clearly fails for the new biology. College is a period of awakening; most students change career interests and intended majors, especially during the first two years. Lower disciplinary boundaries facilitate this exploration. So too does teaching that is investigative, experiential, and connected (see sidebar).

Teaching for a multi-, inter- or non-disciplinary world will require reshaping education at all levels. Current faculty have been trained in disciplines, not across disciplines. Although deep disciplinary knowledge has been a strength of twentieth century science that focused on fundamentals (e.g., physics and chemistry), such narrowness is a handicap now that the frontier of science has

[30] *Greater Expectations: A New Vision of Learning as a Nation Goes to College,* Association of American Colleges and Universities, Washington, DC, 2002.

[31] Bernard L. Madison and Lynn Arthur Steen, *Quantitative Literacy: Why Numeracy Matters for Schools and Colleges,* National Council on Education and the Disciplines, Princeton, NJ, 2003.

[32] www.math.dartmouth.edu/~mqed/index.html

[33] www.css.tayloru.edu/~mdelong/qlsigmaa/frames.html

[34] Elaine Seymour and Nancy M. Hewitt, *Talking About Leaving: Why Undergraduates Leave the Sciences,* Westview Press, Boulder, CO, 1997.

moved on to integrative areas such as genomics, neuroscience, and ecology. Disciplines establish distinctive vocabularies, procedures, and standards for truth. They also build budgets and bureaucracies, both on campus and in the external worlds of policy and funding. If biology is to thrive as the interdisciplinary science that it has become, all these impediments will have to be overcome, if not eliminated. Sidebars throughout this paper illustrate how different organizations are addressing this important challenge. In addition, the National Research Council has begun summer workshops in support of *Bio 2010*[35] that will lend increased credibility to the effort to break down disciplinary silos.

The struggle to integrate the biological and mathematical sciences in students' education faces impediments that are arguably even greater than the general challenge of interdisciplinary study. As noted above, mathematics and statistics have long histories of modeling biological processes, but equally long traditions of separation in education. Because mathematical models are difficult for both students and instructors, because many were (as we have also noted) of limited effectiveness, and because so much biology could be taught by taxonomic and wet-lab methods without significant quantitative challenges, biology came to be seen by everyone as the least mathematical science. Once perceived this way, self-selection

made it that way. Students entered the life sciences in part to limit their exposure to advanced mathematics. When they became professionals, the example of their own career and, often, their advice to students continued this tradition.

This era has passed. Whether they study molecules, cells, or ecosystems, future biologists will clearly need to understand and use sophisticated quantitative and computational tools. So too will anyone dealing with the societal impact of the new biology (genetically engineered crops, epidemics, antibiotic-resistant pathogens, bioterrorism,…). This includes every college student, not only future biologists. It is as important that ordinary citizens understand the new biology as that specialists do. Citizens who elect legislators, police officers who deal with threats, business leaders who make economic decisions, and school board members who set educational policy all need sound understanding of twenty-first century biology. In the new biology, evidence is as often mathematical as observational, as often quantitative as descriptive. In 2010, mathematics and biology will be just as entwined as were mathematics and physics in 1910. We may then marvel, as Wigner did in an earlier era, at the unreasonable effectiveness of mathematics as a tool for explaining how living things function.

---

[35] William Wood and James Gentile, "Meeting Report: The First National Academies Summer Institute for Undergraduate Education in Biology." *Cell Biology Education,* 2 (Winter 2003) 207–209.

A Quantitative Approach to the Biology Curriculum: Issues To Consider

Robert Yuan
University of Maryland-College Park

Robert T. Yuan is Professor of Cell Biology and Molecular Genetics at the University of Maryland-College Park. His research on the molecular biology of bacteria led to the discovery of the first restriction enzymes, an essential tool in genetic engineering. Yuan received his PhD in Molecular Biology from the Albert Einstein College of Medicine in 1966. In addition to molecular biology, his research interests include the use of biotechnology in economic development, traditional Chinese medicines, and development of interdisciplinary and cross-cultural biology courses. Yuan serves as a consultant for the Board of Life Sciences of the National Research Council working on undergraduate biology education, including the report *Bio 2010*. *E-mail*: ry11@umail.umd.edu.

Defining the Challenges

Scientific research is undergoing a major transition from disciplinary approaches to interdisciplinary work. The nature of this momentous shift is examined in *Bio 2010: Transforming Undergraduate Education for Future Research Biologists*, a National Research Council report carried out for the National Institutes of Health and the Howard Hughes Medical Institute (NRC, 2003). A principal goal of the study was to define how the knowledge of mathematics, chemistry, physics, computer science and engineering will enable students to pursue research careers in biology which are increasingly interdisciplinary in nature. The study was carried out by an NRC committee with the collaboration of panels in several disciplines.

"Biology majors headed for research careers need to be educated in a more quantitative manner," reported the study's panel on mathematics and computer sciences. Moreover, they noted, this kind of quantitative education "may require the development of new types of courses." The panel identified certain basic concepts that it deemed essential for such students (e.g., limits, rates of change, computer modeling). The objective of increased quantitative education would be accomplished, they recommended, by introducing quantitative concepts and methods into biology courses while at the same time making biological concepts more prominent parts of courses in the mathematical and computer sciences. They also stressed the importance of fundamentally restructuring introductory biology courses since these courses play a major role in shaping student attitudes towards interdisciplinary concepts.

While *Bio 2010* addressed the needs of a specific group of students, namely those headed for graduate or medical school and research careers, its conclusions are nevertheless relevant to all biology students. The aptly named conference on "Meeting the Challenges" represents a further step in finding practical ways to incorporate quantitative concepts and skills into the biology curriculum.

The task of creating a more quantitative biology curriculum faces a number of distinctive challenges:

- *Students* are becoming increasingly diverse, as are the career tracks for biology graduates.
- *Curricula* to meet this need are not in place. Indeed, the kinds of mathematical knowledge biology students need for their future careers needs to be better ascertained before this knowledge and these skills can be integrated into existing courses or used to create new courses.

- *Faculty* both in biology and in the mathematical and computer sciences will be required to develop and teach new courses. This will require access to the best course models available, but much more as well, namely a change in the academic culture to support major initiatives in faculty development ("teaching the teachers").

- *Institutions* also need to provide support. While faculty may create new courses (often with federal and foundation funding), it is questionable whether such changes are sustainable and can be disseminated to other institutions. The departmental structure of institutions often hinders educational innovation and the system of recognition and rewards most often does not provide incentives for such creative educational effort.

Students: More Diverse Backgrounds and Expanding Career Goals

All faculty members approach courses they teach with a combination of familiarity and precedent— familiarity with their areas of research and the precedent of how each course has been taught in the past. Biology faculty have the additional pressure of a significant number of pre-med students and the topics that will be tested in the MCAT examinations. This is altogether not a promising way in which to approach the teaching of a rapidly evolving, highly interdisciplinary field of study. It is probably more useful to consider two of the major trends in undergraduate science education:

- *The ever-increasing diversity of the student body:* Students in many of our colleges and universities are increasingly diverse in terms of gender, race, ethnicity, age, educational background and career interests.

- *Multiple career tracks:* In the past two decades, we have not only witnessed the emergence of totally new areas in the biological sciences but also seen this new knowledge quickly applied in many different ways.

A striking example of rapid and diverse application can be seen in the new field of genomics, the study and analysis of the DNA in living organisms. In a research laboratory, genomics can be used to study pathogenesis, the process by which microbes cause disease in host organisms. Placed in a different context, it can be used in a museum to study the evolution of plant species. In a police department, it can be applied to forensics. And in a venture capital fund it can be used to evaluate the prospects for a new biotechnology company.

The traditional approach to introducing undergraduates to interdisciplinary work is through research in college or university laboratories. While working under the supervision of a faculty mentor is a superb opportunity, only a small percentage of students ever have this opportunity—and it usually comes in the later part of their undergraduate experience. Consequently, the fundamental solution to the need for all biology students to experience an interdisciplinary approach must lie in a major restructuring of courses in biology and supporting sciences. The introduction of quantitative and computer skills into biology must be comprehensive, starting with the large enrollment introductory courses and progressing all the way into special-topics senior courses.

An important study (Seymour & Hewitt, 1997) showed that the most serious attrition of potential science majors occurs during the first undergraduate years. When surveyed, students who dropped out of their science majors gave as their principal reasons poor preparation, poor teaching, and lack of apparent relevance to potential careers. The creation of a quantitatively-oriented biology curriculum must take into account the broad diversity of both students and their career interests. This can be best done by considering different levels of mathematical knowledge:

- *Literacy:* the ability of students to understand quantification as it relates to some of the principal concepts and experiments in introductory biology courses.

- *Competence:* the ability to acquire the necessary tools that are needed in designing experiments and analyzing increasingly complex sets of data in upper level courses.

- *Expertise:* the ability to apply sophisticated quantitative methods in new biological fields such as ecological simulations and genomics. This most often occurs in graduate school.

Since the design and implementation of quantitative modules and interdisciplinary courses is costly in terms of resources and faculty time, it is of the utmost importance that we have a better understanding of actual needs. Surveys of students and recent graduates in physics and chemistry have provided a good picture of the types of knowledge and skills required for different career tracks—whether those be in research, industry, or non-laboratory related occupations (AIP, 2002). This type of needs assessment could provide the basis for a rational design of quantitatively based biology courses at different levels of the curriculum.

Joint Programs: Some Examples

- *Morgan State University* combines biology with computer science.
- *Goucher College* offers a 3/2 engineering program where students do a three year major in biology and then do two years in engineering at Johns Hopkins University.
- *Cornell University* offers an undergraduate program in mathematics with a concentration in mathematical biology requiring three courses at the interface between mathematics and biology and two additional courses in mathematics.

Recommendation 1. *There should be an assessment of the quantitative needs of biology graduates.*

Curriculum: Creating New Educational Constructs

Changes in the biology curriculum must take into consideration certain constraints:

- 50% or more of biology students are pre-med.
- College and university administrations are unlikely to accept changes that lengthen the degree program or that lead to substantial cost increases.
- Demands for added faculty time or increased numbers of graduate teaching assistants are not likely to be met.

Developing an appropriate educational strategy involves both course design and the availability of teaching tools and resources. Course design includes the following approaches:

- Course modules that integrate quantification with major biological concepts. (Unfortunately, this option may not be practical in that biology faculty are already struggling with the ever-increasing amount of biology they feel students need.)
- Quantitative biology courses that focus on topics that are intrinsically mathematical in nature (e.g., enzyme kinetics).
- Biology-oriented mathematics courses that select examples primarily from the life sciences. (Specialized mathematics courses and quantitative biology courses seem to be the most popular strategies.)

- Supplementary mathematics courses, typically 1 or 2 credit, that provide quantitative support for an existing biology course.
- Parallel courses that involve two versions of a specific course, one being mostly qualitative with its counterpart being more quantitative.
- Special quantitative tracks that typically take the form of a major in quantitative biology that requires several quantitatively focused courses.
- Math Centers that help students acquire mathematical or quantitative literacy. These are often associated with Mathematics Across the Curriculum (MAC) programs.
- Joint programs that combine biology with computer sciences or engineering (see the above sidebar). Such programs typically take an extra year.

These approaches highlight two separate problems: what kinds of courses and programs are needed (or available) and what kinds of tools are needed (or available) to support the courses? The suitability of one or another approach on a particular campus will depend on both the needs and available support tools.

Unquestionably, the single most daunting challenge is the introductory courses. There are already intense controversies over the content of these courses, to which are added the logistic problems arising from large enrollments (e.g., team teaching, multiple sections, teaching assistants, student evaluation). Adding quantification will only make these problems even harder to resolve.

Another major hurdle is the dearth of textbooks and teaching materials that combine mathematics and biology. There are a few examples (see sidebar below), but some of

Mathematics and Biology References

Edelstein-Keshet, L., *Mathematical Models in Biology*. Birkhäuser, 1988.
Hoppensteadt, F.C. and C.S. Peskin, *Mathematics in Medicine and the Life Sciences*. Springer, 1992.
Murray, J.D., *Mathematical Biology*, Springer-Verlag, 1993.
Wood, W.B., J.H. Wilson, R.M. Benbow, and L.E. Hood, *Biochemistry: A Problems Approach*, W.A. Benjamin, Menlo Park, CA, 1974.

those that were best able to provide students with problem solving skills are no longer in print. The lack of appropriate teaching materials represents a major obstacle to the creation of quantitatively oriented biology courses.

Recommendation 2. *Model courses and approaches in quantitative biology along with supporting teaching materials and assessments should be identified and disseminated.*

Faculty: Learning New Ways

Like a number of other studies, *Bio 2010* shows that much innovative teaching takes place in liberal arts colleges and teaching institutions. In contrast, research universities are in the unique position to both generate and disseminate new knowledge. The challenges to the faculty in both kinds of institutions are manifold: introducing new research findings into undergraduate courses, establishing interdisciplinary courses, using faculty time efficiently, creating mechanisms for faculty development, and devising appropriate systems of recognition and rewards for innovative teaching.

Students learn better if their courses present them with exciting new material and provide them with research experiences. If this weren't challenge enough, faculty face the additional problem that teaching is not perceived as being intellectually challenging in the manner that research is. NSF has approached this problem by introducing pilot research programs that attempt to integrate education and research. For its part, the NRC organized a workshop entitled "Integrating Research and Education: Biocomplexity Investigators Explore the Possibilities" that brought together scientists with biocomplexity grants and faculty involved in the design and execution of educational and outreach activities. The objective was to introduce participants to a broad range of educational approaches and encourage them to apply them to their own projects. The credibility of this effort was based on the fact that some of the most attractive

models were devised by individuals who were well recognized for their research.

Using modules or creating new courses or sets of courses is highly demanding in terms of time and resources. It often requires a complex strategy with multiple components:

- Collaboration across departments
- Use of information technology
- Coordination of teaching assistants (both graduate and undergraduate)
- Creation of teaching modules

Such efforts always require more time than does the traditional biology course. Certain new initiatives, however, offer tools to facilitate the work of the faculty. These include:

- On-line teaching materials such as those available on Bioquest (see Sidebar below)
- Postdoctoral fellowships for teaching
- Faculty development summer institutes (see sidebar on facing page)

Bio 2010 recommends faculty development institutes where scientists can be exposed to novel interdisciplinary courses and have an opportunity to develop courses and materials to be used in their own institutions. An exciting new initiative has been the creation of the NRC Summer Institute on Undergraduate Biology Education in collaboration with the University of Wisconsin. This is a summer institute designed along the lines of the famous Cold Spring Harbor courses. Launched in the summer of 2003, it brought together faculty from research universities and teaching colleges and universities, focusing its first workshop on introductory biology courses. The purpose of the institute is faculty development through a process of hands-on experience with successful course models (see facing sidebar). The Mathematical Biosciences Institute (MBI) at The Ohio State University provides similar opportunities that are focused more at graduate students and researchers (see the sidebar on p. 32).

BioQUEST
Beloit College

The BioQUEST Curriculum Consortium, established in 1986, seeks to promote curriculum innovation in undergraduate biology by serving as a national networking resource for individuals to share, distribute, and enhance cooperation among current and future biology education development projects. BioQUEST stands for Quality Undergraduate Education Simulations and Tools in Biology. It focuses reform on the "3Ps" of science education: Problem-posing, Problem-solving, and Peer persuasion. BioQUEST runs faculty workshops, develops software, and has created an extensive index of resources to support undergraduate biology instruction.   bioquest.org

NRC Summer Institute on Undergraduate Biology Education
University of Wisconsin

In addition to calling for changes in the undergraduate education of biologists, *Bio 2010* suggested establishing summer workshops to help implement reform. With enthusiastic support from the National Research Council, the first such workshop was hosted in the summer of 2003 by the Center for Biology Education at the University of Wisconsin, Madison. Inspired by the success of the Cold Spring Harbor Laboratory Courses on specific research areas in molecular biology and genetics, Institute leaders decided to test the idea that the intellectual intensity and excitement of a Cold Spring Harbor course could be generated among participants in a workshop on life sciences pedagogy.

In this Institute, "student" participants were primarily junior biology faculty from Research I institutions, while "instructors" were accomplished biological scientists who had also distinguished themselves as educators. Students and instructors spent days together in intensive lectures, seminars, and laboratory work, but with the goal of learning about new research-based ways to improve undergraduate biology teaching at large universities rather than learning about a new area of biological research.

The pilot Institute was designed around several principles:

- teachers, like students, learn by doing;
- teachers must model the teaching practices we teach about;
- scientists like to be involved in applying new concepts; and
- educators like to leave a meeting with useful tools, not simply new ideas.

Consequently, much of the workshop consisted of sessions in which individuals or teams of participants modeled their favorite simple participatory laboratory or classroom exercises, with the entire group acting as students. These presentations were followed by lively discussion of whether and why the exercise was effective as a teaching tool, and how this tool could be adapted at other institutions.

Further Information:
National Research Council : www.academiessummerinstitute.org/
William Wood and James Gentile, "Meeting Report: The First National Academies Summer Institute for Undergraduate Education in Biology," *Cell Biology Education, 2* (Winter 2003) 207–209. [www.cellbioed.org/articles/vol2no4/article.cfm?articleID=74; www.pubmedcentral.nih.gov/articlerender.fcgi?artid=256984]

www.wisc.edu/cbe/workshops/si/

For many reasons, faculty members are often reluctant to invest more of their time in preparing for courses. How can more extensive learning objectives be met without unrealistic additions on faculty time? As indicated above, the lack of modern interdisciplinary teaching materials is a major obstacle to the creation of new courses. Larger and more expensive textbooks are not a practical solution. Thus there is an ever-growing need for peer reviewed web-based teaching materials. Another powerful tool would be teaching teams that include faculty from different backgrounds, teaching postdoctoral fellows and teaching assistants (who may be either graduate or undergraduate students). These considerations imply several recommendations:

Recommendation 3. *Develop approaches for integrating educational components into research grants.*

Recommendation 4. *Establish faculty development programs to familiarize faculty with interdisciplinary approaches to teaching and enable them to modify or create new courses for their own institutions.*

Recommendation 5. *Develop networks of faculty that can create peer reviewed web-based teaching materials.*

Recommendation 6. *Find more effective ways to use faculty time including use of teaching postdoctoral fellows, graduate and undergraduate teaching assistants.*

Institutions: The Challenge of Sustainable Change

Undergraduate biology education has been undergoing major changes in recent years. Faculty members have

The Mathematical Biosciences Institute
The Ohio State University

The Mathematical Biosciences Institute (MBI) addresses needs that arise from the revolutionary advances in basic science and technology including medical imaging, nanoscale bioengineering, and gene expression arrays. The resulting deluge of experimental data has challenged scientists to produce mathematical solutions to analyzing and structuring this data in a meaningful way.

To address these needs, MBI:

- Develops mathematical theories, statistical methods, and computational algorithms for the solution of fundamental problems in the biosciences;
- Involves mathematical scientists and bioscientists in the solutions of these problems;
- Nurtures a community of scholars through education and support of students and researchers in mathematical biosciences.

MBI's vigorous programs of research and education foster the growth of an international community of researchers in this new field.

In particular, MBI's summer program engages participants in both mathematical and biological approaches to current research. Recent topics from mathematics include oscillations and reaction-diffusion equations, while in biology they include photoreceptor physiology, physiology of synapses and muscle physiology. The 02-03 lecture/workshop series was devoted to mathematical neuroscience; the theme for 03-04 is modeling cell processes. All lectures and reports are available on line.

mbi.osu.edu

been active in developing new courses and teaching materials, often with support from NSF and HHMI. Venues for the publication of biology education papers have increased (see sidebar below), and disciplinary societies such as the American Society for Microbiology and the American Institute of Biological Sciences have made education part of their regular programs. Nonetheless, major problems still remain as regards dissemination and, particularly, the sustainability of new educational initiatives. The latter is perhaps the most difficult challenge in that it entails major changes in institutional structure.

Institutional change requires change both in culture (which at research universities values research over teaching) and in departmental structures (which are built around individual disciplines). For interdisciplinary courses to succeed in the long term, they require:

- A change in the way that credit for individual courses is assigned to instructors and departments. This is particularly relevant in interdisciplinary courses that are team taught by faculty from different departments.
- Freeing faculty time for course development. At the University of Maryland College Park, for example, various faculty awards provide funds to buy out faculty time for the development of new courses.
- Establishment of university-wide Centers for Teaching Excellence that can provide support in the pedagogy of new courses and a system for course evaluation.
- Establishment of training programs for teaching assistants that provide access to interdisciplinary course material and novel teaching methods.
- Changes in the recognition and rewards for innovative teaching. While it is not uncommon for distinguished teachers to receive a departmental or even a university award, this is typically not factored into decisions on salary or promotion. Excellence in teaching should be weighted in a manner comparable to research in establishing norms for raises and promotion.

It must be emphasized that in the absence of institutional support, efforts of faculty, teaching assistants, government agencies, foundations, and disciplinary societies

Biology Education Periodicals

From the American Society for Microbiology (ASM):

Microbiology Educational Journal (annually)

Focus on Microbiology Education (quarterly)

From the American Institute of Biological Sciences (AIBS):

BioScience

From the American Society for Cell Biology:

Cell Biology Education

will fall short of systemic change as regards interdisciplinary biology courses.

Recommendation 7. *The design of new educational initiatives must consider changes in the administrative framework and involve the active participation of academic administrators.*

From the perspective of a biologist, I firmly believe that the changes discussed above must be built around compelling biological concepts. Furthermore, although the term "interdisciplinary " has been used here in relation to the combination of the mathematical and computer sciences with biology, in reality it pertains also to physics, chemistry and engineering as applied to biology. The real challenge for biology is to replace current requirements in mathematics, physics, and chemistry by interdisciplinary biology courses where all of these areas are woven into a continuous fabric.

Bio 2010 argues that the creation of interdisciplinary curricula requires a comprehensive strategy. These include an understanding of student characteristics and career needs, innovative courses, new teaching materials and technology, faculty development, and institutional change. The complexity of the challenges means that they cannot be resolved by a single university, college, or institution. The challenges will ultimately be met by partnerships that bring together academic institutions, government agencies, and industry.

References

American Institute of Physics, "Survey of Bachelor's Degree Recipients," In *The Early Careers of Physics Bachelors*, American Institute of Physics, Aug. 2002.

National Research Council, *Bio 2010: Transforming Undergraduate Education for Future Research Biologists*, National Academies Press, Washington, DC, 2003.

Seymour, Elaine and Nancy M. Hewitt, *Talking About Leaving: Why Undergraduates Leave the Science,* Westview Press, Boulder, CO, 1997.

Wood, William and James Gentile, "Meeting Report: The First National Academies Summer Institute for Undergraduate Education in Biology," *Cell Biology Education*, Vol 2 (Winter 2003) pp. 207–209. www.cellbioed.org/articles/vol2no4/article.cfm?articleID=74

Visualization Techniques in the New Biology

Maria E. Alvarez
El Paso Community College

Maria E. Alvarez is a professor of Biology at El Paso Community College and director of the Research Initiative for Scientific Enhancement (RISE) program that is part of the Minority Biomedical Research Support (MBRS) program of the National Institute of General Medical Sciences (NIH-NIGMS). Alvarez has conducted research on chemical and microbial contamination of the Rio Grande basin and has been a mentor for hundreds of minority students. In 2003 she received a Faculty Mentor Award from the Society for the Advancement of Chicano and Native American Scientists (SACNAS) and a Mentor Role Model award from Minority Access for her role in helping minority students pursue a biomedical research career "at a level not seen at any other community college." Alvarez received a B.S. and M.S. degree in biology from the University of Texas at El Paso and a PhD in biology from New Mexico State University.
E-mail: MariaA@epcc.edu

Partly because data sets and the tools to use them have not been readily available, the traditional undergraduate biology curriculum has not emphasized quantitative tools. Today, however, computer-based technology has made it possible—and imperative—to introduce quantitative methods throughout biology education.

As an example at the simplest level, all introductory microbiology courses require students to characterize bacterial colonies by direct observation. With inexpensive new technology that can reach students at all levels, these labs can be transformed into quantitative exercises based on digital image acquisition. Bacterial cultures on Petri plates can be digitized using a simple scanner. The resulting image file can be measured using public domain software (NIH Image) to yield a database of growth dynamics. Using spreadsheets such as Excel, students can compare bacterial growth characteristics and graph the results. This adds an important new dimension to student learning.

Similarly, modern visualization tools such as Scanning Probe Microscopy (SPM) can transform traditional microscopic observations into exciting investigative exercises in which students can observe and measure three-dimensional images of cells and their components, and observe dynamic processes such as cell division (ASM, 1999). Although few high schools or undergraduate institutions can afford a scanning-probe microscope, new technology makes it possible for students at all levels to have remote access to this equipment via the Internet and to perform or observe experiments in real time. This is the goal of IN-VSEE (Interactive Nano-Visualization in Science and Engineering), an NSF-funded project at Arizona State University involving a consortium of university and industry scientists and engineers, community college and high school science faculty, and museum educators. IN-VSEE is developing educational tools to prepare students for the nanotechnology revolution, which require understanding the structures and properties of matter on a scale below 1000 nanometers. (An overview of nanotechnology can be found at www.zyvex.com/nano/.) The project has developed a website (invsee.asu.edu) where, at the touch of a button, a number of educational modules that utilize remote-access Scanning Probe Microscopy can be used to generate and analyze live data. A similar project (bugscope.beckman.uiuc.edu) at the University of Illinois allows K–12 students and teachers to remotely operate a scanning electron microscope. Other examples of visualization software can be found in the sidebar.

Visualization Software

NIH Image (rsb.info.nih.gov/nih-image) is a public domain image processing and analysis program for the Macintosh. A free PC version of *Image,* called *Scion Image* for Windows, is available from Scion Corporation (www.scioncorp.com). *Image* can acquire, display, edit, enhance, analyze and animate images. It supports many standard image processing functions, including contrast enhancement, density profiling, smoothing, sharpening, and edge detection. *Image* can be used to measure area, centroid, and perimeter of user defined regions, perform automated particle analysis, measure path lengths and angles.

Visualization Software Links (serc.carleton.edu/introgeo/models/visual/software.html) offers a brief list of links to software for visualization separated into two sections: commercial (fee-based) and free or on-line resources.

Molecular Display & Visualization Software (www.netsci.org/Resources/Software/Modeling/Viewers/) is a "software yellow pages" sponsored by NetSci (Network Science, Inc.) intended to provide a comprehensive resource of software options for displaying molecular structures.

Molecular Visualization Freeware (www.umass.edu/microbio/rasmol/) is a package of tools for looking at macromolecular structure. *Protein Explorer,* (derived from the earlier package *RasMol*) enables a user to rotate a protein or DNA molecule to show its 3D structure while *Chime* animates molecules on the web.

MoluCAD (www.kinematics.com/molucad/) is a full-featured molecular modeling and visualization tool (Windows only). Originally conceived as an aid for students of organic chemistry, MoluCAD quickly evolved into a more powerful tool with premium graphical quality and computational robustness. Novice users are able to quickly generate models, view them from any perspective, create reaction animations, and save all data to disk.

Exposing students to these emerging fields that transcend the boundaries of biology, mathematics, computer science, and technology is of paramount importance if we are to prepare students for the global challenges of the twenty-first century. Practical applications of interdisciplinary advances in visualization are already evident. Telemedicine technology (www.bildanalys.se/Med_tele.htm) now permits personnel in hospitals and laboratories in different locations to discuss endoscopy or pathology images in real-time just as is done with multi-headed microscopes or in small groups observing a large video monitor. This allows on-line consultation among specialists, second opinions, off-line consultation over the globe using the Internet, and support for remote diagnosis.

The potential of visualizing and manipulating molecule-sized units of matter is vividly described by Ralph C. Merkle of Zyvex Corp in IEEE's on-line magazine *Spectrum*:

> Nanotechnology will make us healthy and wealthy though not necessarily wise. In a few decades, this emerging manufacturing technology will let us inexpensively arrange atoms and molecules in most of the ways permitted by physical law. It will let us make supercomputers that fit on the head of a pin and fleets of medical nanorobots smaller than a human cell able to eliminate cancer, infections, clogged arteries, and even old age. People will look back on this era with the same feelings we have toward medieval times—when technology was primitive and almost everyone lived in poverty and died young. Besides computers billions of times more powerful than today's, and new medical capabilities that will heal and cure in cases that are now viewed as utterly hopeless, this new and very precise way of fabricating products will also eliminate the pollution from current manufacturing methods. Molecular manufacturing will make exactly what it is supposed to make, no more and no less, and therefore won't make pollutants.

A convincing state-of-the-art visualization system that allows observation and manipulation of images without the limitations of a screen is 3D HoloProjection™ (www.3dh.net). Objects exist in the viewer's reality and can be manipulated and interacted with. For example, this system allows the visual scalability of a live surgical procedure with the ability to record and play back the session for training. Medical students can remotely view live surgery in 3D; PET/CAT/MRI scans can be viewed and manipulated with dimensional clarity anywhere in the world. Researchers can observe and manipulate atomic and chemical interactions and underwater investigations. The applications in education are unlimited.

One teacher can simultaneously teach classes in real time in an unlimited number of classrooms, anywhere in the world, while each student sees her as if she is right in the classroom.

Changing Curriculum

Curricular change aimed at preparing students for the interdisciplinary nature of emerging fields including nanotechnology, telemedicine, 3-D visualization and bioinformatics must engage faculty at all levels of education from elementary schools through college (both two-and four-year) and research universities. Enhancing mathematics and computer skills of biology majors is an essential part of this effort. Biology programs vary widely in quantitative requirements for the major, with some combination of calculus and statistics being most common. Although some institutions require two semesters of calculus and one of statistics, at many two year colleges there may be no mathematics requirements beyond college algebra.

Several years ago the Mathematical Association of America organized a series of discipline-specific workshops looking at the mathematical needs of various fields (Ganter and Barker, 2004). Their recommendations (www.maa.org/cupm/crafty/focus/cf_biology.html) for biology can be packaged into a two-semester sequence that could be implemented at many institutions. The first semester would cover primarily statistical topics while the second semester would cover an integrated version of calculus and modeling. Special emphasis would be placed on graphical representation of data in a variety of formats—a skill that is fundamentally important for biology students. The relative ordering of modeling, statistics, discrete, and calculus topics may vary depending on priorities at individual institutions. A three-semester option was also proposed as well as a third option that would use modules covering mathematical topics as they apply to biological concepts. (Sidebar A2 offers samples of software packages that support these more graphical and quantitative curricula.)

Of course, what is feasible at one institution may not be practical elsewhere. Some institutions have more constraints than others on designing and implementing new courses. In Texas, for example, two-year public colleges must follow the academic course manual mandated by the Texas Higher Education Coordinating Board that lists all approved and transferable courses. Significant changes in curriculum require approval at the state level, which requires additional time. Other states have similar requirements for curricular change. Flexibility and imagination is essential if curricular renewal is to be effective.

Another avenue that may be followed in a variety of educational settings is to introduce quantitative skills in existing biology courses by using modules that allow students to work with data that is specifically relevant to the course in which they are enrolled. This approach has been developed, tested, and implemented at the University of Tennessee under a series of NSF grants. Details can be found at www.tiem.utk.edu/~gross/quant.lifesci.html.

Other pedagogical approaches include case-based studies that encourage students to make connections to real-world situations. The National Center for Case Study Teaching in Science (ublib.buffalo.edu/libraries/projects/cases/case.html) is a repository for case studies that has been assembled by Clyde Herreid and associates at University of Buffalo. For example, Katayoun Chamany of Eugene Lang College at the New School University developed a case titled "Ninos Desaparecidos" that merges the scientific and mathematical principles of genetic identification with human rights. The case focuses on the disappeared children of Argentina and El

---

Mathematical Software for Science

SciLab (www.scilab.org) is a free scientific software package for numerical computations providing a powerful open computing environment for engineering and scientific applications. Scilab includes hundreds of mathematical functions with the possibility to add interactively programs from various languages (C, Fortran...). It has sophisticated data structures (including lists, polynomials, rational functions, linear systems...), an interpreter, and a high level programming language.

Mathematica (www.wolfram.com) seamlessly integrates a numeric and symbolic computational engine, graphics system, programming language, documentation system, and advanced connectivity to other applications. Its wide range of uses include handling complex symbolic calculations; analyzing and visualizing data; solving algebraic and differential equations and minimization problems both numerically and symbolically; and numerical modeling and simulation of complex biological systems, chemical reactions, and environmental impact studies.

Salvador. This case may spark the interest of Hispanic students and of any student interested in global politics and cutting edge technologies. For details, see ublib.buffalo.edu/libraries/projects/cases/ninos/ninos for the case study and ublib.buffalo.edu/libraries/projects/cases/ninos/ninos_notes.html for commentary.

Helpful Resources

Since many biology students have trouble constructing and interpreting graphical data, the investigative approach can be used to address this difficulty. Investigative exercises steer students away from the cookbook approach allowing them to design their own experiment, collect and analyze data, as well as present and defend their results. These types of exercises should be introduced in the curriculum as early as possible.

A few investigative lab manuals for first year biology are available (Dickey, 1995; Dolphin, 1999). Moreover, the National Science Digital Library (www.nsdl.org) provides excellent links to data sets that can be used in the classroom. Engaging students in using data to address scientific questions has long been an integral aspect of science education. Today's information technology provides many new mechanisms for collecting, manipulating, and aggregating data. In addition, large on-line data repositories provide the opportunity for totally new kinds of student experiences. One NSDL site "Using Data in Undergraduate Science Courses" (serc.carleton.edu/research_education/usingdata/report.html) provides information and discussion for educators interested in effective teaching methods and pedagogical approaches for using data in the classroom.

Relatively few textbooks introduce mathematical concepts to life science students. Causton (1977) pioneered this effort by introducing a textbook that covers mathematical concepts necessary for first and second year biology students. A more recent example is Fusaro and Kenshaw's *Environmental Mathematics in the Classroom* (2003), which covers mathematics concepts that can be used in a general education course as well as challenging chapters on environmental concepts suitable for prospective mathematics majors. Similar books are needed to supplement other biology courses. Two recent additions to this literature are Britton (2003) and Murray (2003). There is a great need for new textbooks or e-books to be developed to incorporate emerging field applications.

Science Education Links

Access Excellence (www.accessexcellence.org): The education arm of the National Health Museum (www.nationalhealthmuseum.org) is designed to educate and motivate people about health and the human body.

BioQuest (bioquest.org): BioQuest is a consortium that supports collaborative development of curricula for undergraduate biology. The site includes a library, web links, notes, and information on workshops.

Bio-Rad (www.bio-rad.org): The Life Science Education tab of the Bio-Rad Laboratories website contains resources to help teachers bridge the gap between science in the real world and science in the classroom.

Biotechnology Information Directory (www.cato.com/biotech/bio-edu.html): A "virtual library" of links to education resources in the field of biotechnology; sponsored by Cato Research, Ltd.

ChemConnections (mc2.cchem.berkeley.edu/): Includes modules and other resources for the first two years of college chemistry.

Human Genome Project (www.doegenomes.org): A comprehensive Department of Energy website on genomics, featuring general information and education resources.

MERLOT (merlot.org): The Multimedia Educational Resource for Learning and Online Teaching is a free resource for faculty and students of higher education that maintains links to online learning materials with annotations such as peer reviews.

Microbes Count (www.bioquest.org:16080/microbescount/): A collection of multimedia resources, simulations, and tools that offer an interactive, open-ended environment for learning microbiology.

National Association of Biology Teachers (www.nabt.org/sup/resources/): Web resources for biology teachers, including in "Biotechnology on a Shoestring," "Microbial Literacy Collaborative," and "Neuroscience Laboratory."

Alternatively, since computer technology makes possible live data acquisition, data analysis, and visualization, software resources can enrich biology courses and laboratories even in the absence of quantitatively-oriented textbooks. Good resources are already available for all educational levels. Three sources have already been mentioned: IN-VSEE (invsee.asu.edu), Bugscope (bugscope.beckman.uiuc.edu), and the National Science Digital Library (www.nsdl.org). Others include BioQuest (bioquest.org), BioScieEdNet or BEN (www.biosciednet.org), and Project Kaleidoscope (www.pkal.org). The sidebar contains examples of others.

Learning about Learning

Many educational researchers (e.g., Lester and William, 2000) have noted the importance of gathering data in order to evaluate the efficacy of various instructional approaches in mathematics education. Similarly, the introduction of innovative pedagogy aimed at enhancing mathematics skills of biology students must be documented and evaluated. Systematic evaluation of teaching strategies applied to different student populations and published in peer-reviewed journals can provide the basis for a "best-practices" resource book.

Both mathematicians and biologists who struggle in college classrooms to teach mathematical concepts in biology courses must take into consideration what mathematics education experts have reported. For example, what works for one student population may not work if universally applied. Cultural factors, access to resources, different learning styles, pre-existing perceptions, and metalinguistic factors—to name just a few—have been shown to affect students' understanding of mathematical concepts. These factors must now be studied, documented and evaluated as applied to biological concepts. Although the majority of studies on the effects of different factors on understanding mathematical concepts have focused on K–12 students, these students bring to college classrooms the same (or perhaps augmented) challenges. A few examples will document the need for careful attention to these issues.

MacGregor and Price (1999) investigated whether three cognitive components of language proficiency—metalinguistic awareness of symbol, syntax, and ambiguity—were associated with students' success in learning the notation of algebra. Pencil-and-paper tests were given to assess students' metalinguistic awareness and their ability to use algebraic notation. In a total sample of more than 1500 students aged 11 to 15, very few students with low metalinguistic awareness scores achieved high algebra scores.

Gutstein et al. (1997) examined mathematics instruction and its intersection with culturally relevant teaching in an elementary/middle school in a Mexican American community. They developed a model of culturally relevant mathematics instruction, which takes into account students' culture as well as both informal and experiential mathematical knowledge.

A classic study (Resnick, 1983) revealed that students do not come to classrooms as "blank slates." Rather, they arrive with a variety of misconceptions that interfere with learning mathematics. A catalogue of basic mathematics misconceptions can be found in Benander and Clement (1985) and also on the web at www.counton.org/resources/misconceptions/index.shtml (part of a web site called "Count On" sponsored by the Department of Education and Employment in the United Kingdom.)

In a follow-up study, Jose Mestre (1989) offers evidence that language differences may cause Hispanics to commit errors with a higher frequency. (A condensation of Mestre's article with a variety of examples appears in the sidebar on p. 40.) Other ethnic groups exhibit similarly unique misconceptions about mathematical language (Cocking and Mestre, 1988).

These examples of educational research illustrate the complexity of the issues and challenges educators face when attempting to reach diverse populations of students. There is no single "magic bullet" that will solve the problems of achieving mathematical literacy across the board. Consequently, it is of paramount importance to conduct and document classroom research on mathematics teaching and learning strategies at the college level.

Educating Faculty

Curricular changes will require re-training of faculty at many institutions. A common and effective way to accomplish this is by means of faculty development workshops organized by professional societies. For example:

- The Mathematical Association of America's PREP (PRofessional Enhancement Program) workshops (www.maa.org/prep/) offer faculty in the mathematical sciences extended professional development experiences. PREP workshops serve faculty in the mathematical sciences from all types of institutions at all stages in their careers.

- The American Society for Microbiology (www.asm.org) holds an annual Conference for Undergraduate

Hispanic and Anglo Students' Misconceptions in Mathematics
Jose Mestre, University of Massachusetts at Amherst

Students do not come to the classroom as "blank slates." Instead, they come with theories constructed from their everyday experiences. They have actively constructed these theories, an activity crucial to all successful learning. Some of the theories that students use to make sense of the world are, however, incomplete half-truths. They are what we call "misconceptions."

Misconceptions are a problem for two reasons. First, they interfere with learning when students use them to interpret new experiences. Second, students are emotionally and intellectually attached to their misconceptions because they have actively constructed them. Hence, students give up their misconceptions only with great reluctance.

These findings suggest that repeating a lesson or making it clearer will not help students who base their reasoning on strongly held misconceptions. In fact, students who overcome a misconception after ordinary instruction often return to it only a short time later.

A very prevalent misconception surfaces in the "students and professors" problem: *Write an equation using the variables S and P to represent the following statement: "There are six times as many students as professors at a certain university." Use "S" for the number of students and "P" for the number of professors.*

The most common error in this problem—committed by about 35% of college engineering majors—is to write "$6S = P$." (The correct equation, of course, is "$S = 6P$.") Students thoughtlessly translate the words of the problem from left to right and confuse the idea of variable and label: they interpret "S" and "P" in the equation as labels (verbal shorthand) for "students" and "professors/" rather than as variables that stand for numerical expressions.

Students often incorrectly calculate the original price from a sale price by applying the discount to the known sale price, instead of to an unknown original price. They often misconceive the independent nature of chance events (after getting heads on four consecutive tosses of a coin, they believe that tails are more likely than heads on later tosses). More surprising, perhaps, is that even some college students want to use subtraction to solve problems such as "Margaret had 2/3 of a gallon of ice cream. She ate 1/4 of it. How much ice cream did she eat?" They expect to use multiplication only for computing increases.

A few studies that have investigated mathematical misconceptions among Hispanic students show that they have some unique difficulties that are nearly always the result of differences in language or culture. In the "students and professors" problem, Hispanics sometimes write the answers, "$6S = 6P$" or "$6S + P = T$." In the former case, they reason that the phrase "as many students as professors" implies an equal number of each. In the latter case, students claim that their equation (in which T = total number of students and professors) combines everyone in the correct proportions. In both cases, misconceptions come from language differences.

"A carpenter bought an equal number of nails and screws for $5.70. If each nail costs $.02 and each screw costs $.03, how many nails and how many screws did he buy?" Some Hispanic students believed that the carpenter spent an equal amount of money on nails and screws. Again, the influence of language is clear. The number of unique errors among Hispanics resulting from linguistic difficulties is, however, small. In general, they cause Hispanics to commit the same types of errors as Anglos, but with a higher frequency.

Adapted and condensed from an ERIC Digest article, Mestre (1989)

Educators (www.asmcue.org) just before their annual meeting.

• The Society for Mathematical Biology (www.smb.org) provides training and web-based resources to teach and highlight the importance of mathematics in biology.

• The National Science Teachers Association (www.nsta.org) website features a variety of educational resources for all educational levels as well as a variety of workshops. A recent example is a workshop for high school teachers, cosponsored with the federal

Food and Drug Adminstration (FDA), on the subject of food safety.

- NSF's Chautauqua workshops (www.chautauqua. pitt.edu/ index.html) offer college teachers a means to keep their teaching current with respect to both content and pedagogy. For example, the sidebar below describes a Chautauqua workshop on using the powerful mathematical package *Mathematica* in science classrooms.

- The Institute for Transforming Undergraduate Education at the University of Delaware (www.udel.edu/inst) provides a variety of faculty development opportunities and resources focusing on problem-based-learning.

These professional development experiences are especially valuable when inter- or intra-institutional as well as interdisciplinary teams of faculty participate.

Another approach, when it can be arranged, is for college and secondary school mathematics faculty to participate in on-going biological research projects that deal with quantitative interdisciplinary topics. For example,

the Dolan DNA Learning Center (www.dnalc.org) provides opportunities for teams of high school and college faculty to receive training in genomics at Cold Spring Harbor Laboratories. The Center also features science education resources and workshops on a variety of topics for students and faculty at all levels. The DOE Laboratory Science Teacher Professional Development program (www.scied.science.doe.gov) allows K–12 teachers and community college faculty to work on interdisciplinary research and education projects. The workshops provide the tools for the participants to become leaders of science education reform. Teachers and college faculty who participate in these learning experiences renew their enthusiasm for teaching, and undoubtedly act as catalysts of curricular reform at their schools and colleges.

With current constraints on institutional budgets, it is essential for professional societies and funding agencies to make the information and training provided during workshops available via the Internet to a larger audience that may not be able to travel to distant meeting sites. As noted above, the National Science Digital Library

Chautauqua Short Courses
Mathematica

Chautauqua Short Courses are an annual series of forums in which scholars at the frontiers of various sciences meet intensively for several days with undergraduate college teachers of science. The series is held at colleges and universities throughout the United States as well as at selected special sites. These forums provide an opportunity for invited scholars to communicate new knowledge, concepts, and techniques directly to college teachers in ways that are immediately beneficial to their teaching. The primary aim is to enable undergraduate teachers in the sciences to keep their teaching current with respect to both content and pedagogy. From year to year, there is an attempt to rotate courses among different regions of the country.

For example, in 2003 one of the workshops was on "Data Analysis and Visualization Using *Mathematica*." It was taught by Flip Phillips, Assistant Professor of Psychology at Skidmore College, a former musician, who is currently editor of *The Mathematica Journal*. From the course description:

This course will address the use of the software *Mathematica* in the science classroom emphasizing graphical presentation techniques. More than just a tool for teaching mathematics, *Mathematica* is a complete scientific computing environment with applications available in a broad range of disciplines, including … statistics, computer science, and the biological and social sciences. In this course we will … conduct a survey of its many uses, including but not limited to technical problem solving, programming, and document preparation and presentation…. Segments will appeal to a wide array of prior *Mathematica* knowledge. Initial sessions will address a series of usage and programming techniques. Subsequently, participants will receive hands-on experience with various discipline specific add-on packages and with the publicly available material from MathSource, the *Mathematica* notebook repository. We will also survey current classroom and teaching laboratory uses of *Mathematica*.

Further information:
Chautauqua Short Courses: www.chautauqua.pitt.edu/index.html
Mathematica: www.wolfram.com

Answering the Question: What is Mathematics Good For?

The Department of Mathematics at the British Columbia Institute of Technology has created a web page (www.math.bcit.ca/examples/index.shtml) intended to help students who want to know "Exactly How Is Mathematics Used In Technology?" The main page consists of a large matrix whose rows are areas of technology (e.g., biomedical engineering, nuclear medicine) and whose columns are areas of mathematics (e.g., algebra and geometry, differential calculus). Clicking on a cell takes you to one or more examples of how a particular area of mathematics is used in a particular area of technology. For example, clicking on the cell for uses of algebra and geometry in nuclear medicine produces the following problem and solution:

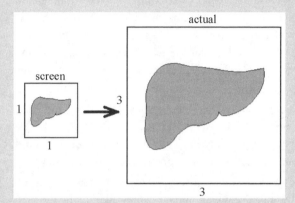

Problem: In a certain imaging study the screen is set to a scale of 1:3 (i.e., 1 cm viewed on the screen is equivalent to 3 cm in the person's body.) If a screen view of the liver displays a lesion area of 1.6 cm^2, how large is the lesion area in the person's liver?

Solution: You might say the solution is an easy ratio and proportion, or is it? The answer is not as simple as 3×1.6. The trick is to recognize that when area is involved both dimensions (length and width) are changing. If you visualize a square 1 unit by 1 unit that expands to a square 3 units by 3 units, the area of the square increases by a factor of 9:

$$\frac{AreaOriginal}{AreaNew} = \frac{1}{3} \times \frac{1}{3} = \left(\frac{1}{3}\right)^2 = \frac{1}{9}.$$

So the correct solution in this case is:

$$\frac{AreaScreen}{AreaBody} = \frac{1.6\,cm^2}{x} = \left(\frac{1}{3}\right)^2 \rightarrow x = 1.6 \times 3^2 = 14\,cm^2.$$

(www.nsdl.org) contains a great deal of information on using data in undergraduate science classes. In addition, the Math Forum at Drexel University (mathforum.org) is an excellent source of practical examples and other discussions concerning the role of mathematics. An example of one of the links available at the Math Forum is illustrated in the sidebar above—answers to the question "What is mathematics good for?"

Web-based conferences as well as local or regional meetings can bring people together at minimal cost. Such meetings can attract faculty from small colleges and high schools, and would also make it feasible to bring mathematics and biology teachers (and students) together for joint workshops. Stimulated by high quality on-line resources, local workshops can provide an interdisciplinary framework to train faculty in new scientific concepts, quantitative methods, pedagogical approaches and curricular examples.

Summary

The relevance of mathematics to the biological sciences (and all scientific fields) must be emphasized as early as possible and on a continuous basis so that students understand the importance of learning mathematical concepts. To meet the challenge of emerging interdisciplinary biological fields, we need to develop a mathematical "culture" that includes students, teachers, and parents at all levels. As it is important to enhance the mathematical skills of biology majors, it is equally important to encourage mathematics and computer science majors to develop interests in biological fields. Students with strong quantitative skills should be actively exposed to emerging and exciting fields of nanotechnology, telemedicine, and bioinformatics to help them see the value of an interdisciplinary education.

The importance of these new fields combined with the

immense complexity of problems they encompass requires extraordinary effort. For students, it means career counseling, scholarships, grants, awards at science fairs, and research internships as early as possible. For faculty, it requires focused workshops on teaching strategies and grant funding, accessible websites, downloadable tools, and interdisciplinary conferences. And for the public, it is imperative to develop and maintain a media campaign to inform all audiences about the potential of these emerging fields, the importance of mathematics to their development, and more generally, the value of an interdisciplinary education. One special area of concern are populations that have been traditionally underrepresented in science, mathematics and engineering fields such as women and minorities. Focused efforts with an inherent sensitivity and understanding of the cultural, social and economic factors that impact underrepresented groups are essential.

References

ASM, "Seeing is Believing: Impact of New Scanning-Probe Microscopes," *Focus on Microbiology Education,* Fall 1999, American Society for Microbiology, Washington, DC.

Benander, L., and J. Clement, "Catalogue of Error Patterns Observed in Courses on Basic Mathematics," University of Massachusetts, Amherst, Scientific Reasoning Research Institute, Hasbrouck Laboratory, 1985. (ERIC Document Reproduction Service No. ED 287 762)

Britton, Nicholas F., *Essential Mathematical Biology*, Springer-Verlag, New York, 2003.

Cocking, R., and J. Mestre, (eds.) *Linguistic and Cultural Influences on Learning Mathematics,* Lawrence Erlbaum Associates, Hillsdale, NJ, 1988.

Causton, David R., *A Biologist's Mathematics,* Edward Arnold Publishers, London, 1977.

Dickey, Jean, *Laboratory Investigations for Biology*, Benjamin Cummings, Redwood City, CA, 1995.

Dolphin, Warren D., *Biological Investigations,* Fifth Ed., WCB/McGraw-Hill, New York, 1999.

Fusaro, B.A., and P.C. Kenshaft, (eds.), *Environmental Mathematics in the Classroom,* The Mathematical Association of America, Washington, DC, 2003.

Ganter, Susan and William Barker, (eds.) *Curriculum Foundations Project: Voices of the Partner Disciplines.,* The Mathematical Association of America, Washington, DC, 2004.

Gutstein, Eric, Pauline Lipman, Patricia Hernandez and Rebeca de los Reyes. "Culturally Relevant Mathematics Teaching in a Mexican American Context," *Journal for Research in Mathematics Education* 28:6 (December 1997) 709–737.

Lester, Frank K. and Dylan William, "The Evidential Basis for Knowledge Claims in Mathematics Education Research," *Journal for Research in Mathematics Education* 31:2 (March 2000) 132–137.

MacGregor, Mollie and Elizabeth Price, "An Exploration of Aspects of Language Proficiency and Algebra Learning," *Journal for Research in Mathematics Education* 30:4 (July 1999) 449–467.

Merkle, Ralph C., "Nanotechnology: What Will it Mean?" Spectrum OnLine. IEEE, www.spectrum.ieee.org/WEBONLY/resource/speakm.html.

Mestre, Jose, "Hispanic and Anglo Students' Misconceptions in Mathematics," ERIC Digest ED313192, 1989. Web: www.ericfacility.net/databases/ERIC_Digests/ed313192.html.

Murray, J.D., *Mathematical Biology I, II*, Springer-Verlag, New York, 2003.

Resnick, L., "Mathematics and science learning: A new conception," *Science* 220 (1983) 477–478.

Building the Renaissance Team

Carol A. Brewer
University of Montana

and

Daniel Maki
Indiana University

Carol A. Brewer is Associate Professor in the Division of Biological Sciences at the University of Montana where she carries on research programs both in physiological ecology and in science education. The former is focused on functional morphology of leaves and conservation biology of temperate forests in southern South America. Brewer received her PhD from the University of Wyoming in 1993. She is currently a member of the Committee for K–12 Education of the National Research Council and Vice President for Education and Human Resources of the Ecological Society of America.
E-mail: carol.brewer@umontana.edu.

Daniel Maki is Professor of Mathematics at Indiana University and author of several textbooks on finite mathematics and mathematical modeling. At Indiana he has served as director of graduate studies and department chair; he also directed NSF-sponsored projects on Research Experiences for Undergraduates (REU) and Mathematics and Its Applications Throughout the Curriculum (MATC). Maki received his PhD from the University of Michigan in 1966. He is currently a member of the Board of Governors and Executive Committee of the Mathematical Association of America. In 2004 he received the President's Award at Indiana University for his leadership in project-oriented education. E-mail: maki@indiana.edu.

Recent research using tools from the biological, mathematical and computer sciences has led to dramatic improvement in our understanding of biology, medicine, and the environment. New fields such as bioinformatics and data mining combined with powerful new computational tools are answering important biological questions. They are also generating new questions, thus defining the frontier of research in the life sciences. As these new cross-disciplinary fields continue to develop new knowledge, techniques, and processes, they create even more opportunities for new biological research.

To cite one example among many, biologists were able to work out the structure of proteins by using a mixture of rigorous strategies (e.g., dynamic programming) and heuristic methods to align sequences. This led to the discovery of domains in proteins—discrete portions of a protein with their own function. Subsequently, biologists discovered that protein sequences are conserved over vast evolutionary distances (e.g., Thornton and DeSalle, 2000). One result has been a growth industry in developing and using methods to align sequences.

If the potential of these exciting new avenues of research is to be realized, a new generation of scientists will be needed. Clearly if biologists, mathematicians, and computer scientists are to work together to solve complex biological questions, more diverse training will be needed at all stages of education. Of course, it is impossible to know in advance which students will choose careers in fields such as biology, biotechnology, or medicine. Thus we need to view all undergraduate biology, computer science, and mathematics majors as potential researchers in biology and the medical sciences. To accomplish this, educational programs need to be redesigned to prepare students to work in the kinds of cross-disciplinary teams required by the "new biology" (Brewer and Gross, 2003).

Although it is impossible to forecast the future, we can safely predict that most students will change careers several times during their lives and that those who do stay in the same career will often find that the nature of their work has changed. So the issue for educators (and their students) becomes: "What skills will the next generation of life scientists need? What can educators do now to ensure that their graduates have the skills they need for future careers in the life sciences?" In this short paper, we will look at what skills students might need in the future, what an ideal working team might look like, and what barriers need to be addressed to achieve these goals. At the end we offer one example of an integrated curriculum designed to develop the skills we believe students need.

Scientists, mathematicians, and educators poised at the beginning of a new century of discovery in the biological sciences need to fully investigate these new questions and not be satisfied with old answers. How we mobilize today to meet the challenges of the future will determine how well prepared our graduates will be to explore and participate in these new intellectual frontiers.

What Students Need

Regardless of their interests, pre-college preparation, declared majors, or career goals, if undergraduates are to be competitive in an ever-changing work environment they must be prepared to adapt and learn new material quickly. Students must learn how to think critically, to evaluate evidence, and to solve open-ended problems. Ideally, they should also have experience working in teams on problems that are inherently interdisciplinary. To this end, all students should have opportunities to take courses that blend theory and practice and provide research-like activities and projects. In addition, courses should focus on process supported by cogent content examples and plenty of open-ended problems.

These general goals are particularly relevant to undergraduate biology, mathematics, or computer science curricula. Even though many students in these subjects do not end up working in their major field (however broadly we define it), much less in biological or medical research, colleges still must prepare all students to be critical thinkers and scientifically literate citizens. Moreover, regardless of where their career paths might eventually lead them, at some point in their careers most graduates will work on a cross-disciplinary team. Therefore, we need to prepare all majors to work together and to view their disciplines as interconnected. In the words of the National Research Council's report *Bio 2010* (NRC, 2003), it is necessary that "interdisciplinary thinking and work become second nature."

In the past, those who pursued careers in biology or medicine rarely studied mathematics or computer science in depth; in fact, many were mathematics- and technology-averse. Today, however, it is nearly impossible for a student to pursue a career in life science research without having a strong foundation in both of these fields. In addition, whether investigators are focusing on evolutionary biology, genomics, or environmental change, it is unlikely that a single individual will know everything necessary to pursue these new avenues of

research alone. More likely, research will be conducted by multi-disciplinary teams to which biologists, mathematicians, and computer scientists all contribute. This is true equally in both basic research (e.g., decoding the DNA of a virus) and applied investigations (e.g., monitoring the spread of SARS). For example, creating a flu vaccine each year requires analysis of data from prior years, probabilistic estimates of various paths in the evolution of the virus, the economics of manufacturing and distribution, and the politics of possible misjudgments in the seriousness of the threat—a truly interdisciplinary exercise.

Regardless of the disciplines represented, interdisciplinary teams work best if everyone has some knowledge of the big picture and some understanding of what each member of the team is contributing. Thus, whether or not they find themselves in a research environment, those working in biology or biotechnology need to understand and be able to use mathematics and computer science. For the same reasons, mathematicians and computer science need to know enough biology to understand the key questions so that they can help develop the tools required to solve these problems. College faculty need to examine seriously whether undergraduates, regardless of their career paths, are being prepared to participate in a multi-disciplinary world where teamwork and sharing expertise is becoming less the exception and more the rule.

Seeking New Foundations

Educators' traditional response to what students need to know is to simply list the courses they should take during their undergraduate studies and the general topics which they should be exposed to. But what does "exposed to" mean when setting up an actual curriculum with real courses and syllabi? Does it mean providing superficial coverage or in-depth opportunities to explore the topics? Does it mean being prepared for the next course in the sequence or for more work in the field, while also encouraging the student to want to work in the field?

For example, biology students of the future will need a strong foundation in probability theory and stochastic processes. Ecology, genetics, and medical testing cannot be understood without understanding statistics and probability theory. The same is true for epidemiology, which started with deterministic models but has since moved on to using more stochastic modeling. Mathematics and computer science students, on the other hand, will need a good foundation in biological principles along with a basic understanding of some of the questions at the frontier of modern biology in areas such as genomics, evolutionary biology, and biocomplexity.

Within the current curriculum the only way students can prepare for the cross-disciplinary needs of diverse careers in the biological and health sciences is to major in one area of science or mathematics and take as many classes as possible in another area. Computer and mathematics students might take classes in the basics of biology, chemistry, and physics, while biology students, might include classes in calculus, probability theory, linear algebra, and databases. But for almost all students this option is too inefficient and expensive since it extends an already full program beyond what can reasonably by accomplished in a standard four year undergraduate program.

Instead, we propose an alternative approach that places students at the center of the educational enterprise and defines education by means of outcomes rather than time or credits. Rather than ask what classes should students take, we work backwards from what we would like our students to be able to do.

The Ideal Cross-Disciplinary Team

When approaching any interdisciplinary question, members of a research team are not necessarily experts in all relevant areas. In fact, rarely are team members strong in all areas of concern. Rather, they bring to the question areas of strength and an ability to communicate and collaborate with others with complementary training and interests. To help define what this means for students, we've identified four areas required for members for our renaissance team: content expertise, shared knowledge, communication skills, and teamwork.

Content Expertise. To be a contributing member of a team, all members need at minimum in-depth applied expertise in one critical area of the investigation, be it ecological diversity, population biology, modeling environmental change, or visualizing large sets of data over time (e.g., population densities). Such expertise would include having participated as a student in research projects in the area of study. Abstract knowledge of modeling techniques and traditional models is not enough. Instead, students need experience working in a real world setting, collecting and examining data, attempting to fit models to those data, and trying to learn from the models.

Shared Knowledge. As noted above, no single member of a team can be expected to have expertise in all relevant areas. It is expected, however, that all members of the team have a strong foundation in the general areas under investigation. In fields where they are not experts, team members should be able to ask good questions and read in other disciplines in order to better understand the problems in that field.

While a mathematician or computer scientist might not know specific details about a specific biological situation (for example, biological diversity of salmon), she or he should understand the basic principles of population biology and genetic diversity, and the critical role rivers play in fish spawning. On the other hand, a fish biologist working on salmon recovery might not know how to model population diversity including influences such as weather, genetic variability, or the potential breaching of dams, but he or she should know that these are critical factors for constructing robust models. The biologist needs to be able to critique proposed models for predicting how changes in one factor might affect salmon recovery, including the assumptions upon which the models are based. Computer scientists, on the other hand, must understand the models well enough to be able to extract biologically useful information.

Communication Skills. Too often, scientists and mathematicians do not have a common vocabulary for communicating basic concepts. Ideal team members need the ability to understand each other and to teach each other using familiar terms. They ask good questions and maintain a conversation from a variety of perspectives to advance the common knowledge and to ensure that everyone contributes to solving the problem at hand. In other words, effective team members need to be willing and able to find common ground by breaking down the barriers that separate areas of expertise.

Students (future team members) also need to be able to communicate clearly and succinctly in a variety of formats. They need to be able to report the results of their work in writing at different levels—not only to their peers, but also to non-specialists and non-scientists. Writing skills should be complemented with the ability to communicate data both orally and visually. Presentations enhance communication and often set the framework for more detailed written reports. They frequently require a variety of visual aids that require experience to use effectively. Finally, students need to learn how to match the technical details of a presentation with the ability of the audience to grasp such details. There is no substitute for practice: students need to give presentations early and often.

Teamwork. Teamwork means more than just two or more investigators working side by side on different aspects of a problem. Rather, research teams will be composed of those who can bring their expertise to the problem, communicate with one another about that expertise, and work together on moving a problem forward. Clearly, team researchers do not have to know every fact or every discipline in depth, but they will need to know how to actively engage the expertise of the others each step along the way. One of the key benefits of having students work in teams is to allow them the opportunity to learn to be a team player in an academic setting. They must be able to cooperate, contribute, compromise, and criticize in a way that helps the team.

These are the kinds of skills that will enable individuals with diverse backgrounds and training to collaborate, with each member contributing his/her expertise to a common goal. In the direction the world of biology is heading, where questions of increasing complexity cannot be addressed except by teams with multi-disciplinary skills, the market-place value of the lone individual who solves a problem on his or her own will decrease, while the value of those who are able to work with others on these new kinds of multi-disciplinary questions will increase.

Designing A Renaissance Campus

Assuming we want students to be able to join our renaissance teams as full participants, what academic skills should they develop during their tenure on campus? As importantly, what experiences should they have to equip them for the world outside the academic setting?

Our goals are clear: to provide undergraduates in biology, mathematics, and computer science with opportunities (a) to learn deeply in at least one discipline and broadly in one or two others; (b) to develop their abilities to communicate and work closely with their peers from different disciplines; and (c) to experience the challenges and scientific benefits of interdisciplinary learning.

The challenges are also clear: The segregation of disciplines into different departments (even different administrative colleges) is inherent in the structure of universities. As anyone who has tried to establish an interdisciplinary course can confirm, it is exceedingly difficult (although not impossible) to bring faculty together to cooperatively teach a course. It is even more difficult when such a course requires listing in multiple departments because then questions are raised about which faculty member gets credit for the class, which department gets credit for the student enrollments, etc. For that reason (and many others), faculty are often discouraged from organizing or participating in such cross-disciplinary courses.

Yet if the goal of education is to focus on the student, what we really need are administrative structures to make such courses not only possible, but routine. Change can take a variety of forms, but if the goal is for universities to become more student centered—and if as we have argued cross-disciplinary experiences are in the best interests of students—we need to develop programs that help the university better achieve its goals of quality education for all of its students while protecting the intellectual and professional development of faculty (as is now done now through the departmental structure). To this end, we offer several possible alternatives.

First, at the faculty level, all science, mathematics, and computer science faculty need to view their educational practice as more closely aligned with their research practices. Although many faculty members in the sciences routinely collaborate for research within their discipline, fewer collaborate across disciplines, let alone translate that same kind of interdisciplinary environment into their classrooms. Yet if the classroom is not interdisciplinary (or at least open to other disciplines and perspectives), students will learn disciplines in isolation without understanding the depth and breadth of the subject in its real-world context.

To this end, faculty should highlight the unique relationships between disciplines, rather than focus on their distinctions. And, much as they do in their own labs, faculty need to encourage students to collaborate and learn from one another. In many of the sciences, the atmosphere of the lab is one which integrates advanced undergraduates, graduate students, postdocs, and senior researchers. However, the integration is usually only in one field. The next step is to welcome students and researchers from other areas and to let them experience the atmosphere of a team approach.

For example, to develop effective ecological forecasts (see Clark et al., 2001) a team would need expertise in the principles of ecology, probability (e.g., random variables, stochastic processes), Bayesian statistics, numerical analysis, and computational science. Participation in such teams as part of their undergraduate experience would be especially helpful for mathematics and computer science majors. Indeed, faculty must move beyond the way they were taught as undergraduate and graduate students to develop more effective methods of instruction. In many cases, their graduate school experiences could provide better models of the kinds of interactions that foster interdisciplinary teaching and learning.

At the university level, department heads and other administrators need to develop new ways to reward faculty. Many faculty report that they would be willing to participate in interdisciplinary courses and become more involved in their teaching, but too often the reward structures do not properly acknowledge their participation. Thus the threat looms that extra time spent on teaching at the expense of research may be held against them when they come up for tenure. As a result, frequently the only faculty who can afford to spend the time needed to experiment with new methods of teaching and learning are tenured, near-retirement faculty rather than the younger faculty who could sustain these kinds of student-centered changes.

The Renaissance Campus cannot emerge without changes at the national funding level where it is important that research support move beyond the principal investigator model. The National Science Foundation (NSF), the National Institutes of Health (NIH), and other funding agencies should acknowledge in their own granting procedures that the "lone ranger" investigator model does not necessarily work in today's complex research environment and that it often takes more than one investigator to make a project successful. While most funding agencies now seem interested in supporting interdisciplinary projects, researchers need to know that this is a serious and long-term commitment.

In mathematics, for example, funding has a long tradition of single investigator projects. As a result, it is very difficult for mathematicians to be funded as part of an interdisciplinary team. However, many of the large state-of-the-art research initiatives in biology now funded by NSF, NIH, and others are interdisciplinary, requiring the project to be headed by three or four collaborating investigators. These projects might be a more relevant choice for interested mathematicians in the future.

Achieving a More Integrated Curriculum

If these barriers could be reduced to enable the university structure to become more student-centered, what kind of educational structure might better prepare our Renaissance team? Do we need to tear down the entire structure of the campus and build anew? Or can we work within the existing structure and try to improve it?

On our Renaissance campus, lower division students would still take introductory courses in biology and other sciences, in mathematics, and in computer science. But think for a moment what it might be like if those courses were designed not just to cover the material needed to take the next course in the discipline, but rather were designed from the outset to introduce some of the ideas

we've discussed here. Suppose, for example, that all faculty who taught introductory courses in biology, chemistry, and physics met with their colleagues in mathematics and computer sciences and identified a common theme or two (e.g., global warming or Sudden Acute Respiratory Syndrome (SARS)) that they could introduce into one or two sections of both their lectures and labs? Such themes would emphasize the interdisciplinary nature of science and introduce early on many of the mathematical, computer, and communication skills we have been emphasizing. As students progress through this kind of sequence, they might have opportunities to share labs with students in another science sequence, and together be required to gather data and, drawing on their written, verbal and visual communication skills, make a presentation to their peers.

We also can envision introducing complex real-world problems of local interest in the last quarter or semester of large survey courses. For example, Montana has been subjected to a large number of wildfires, which has intensified the debate between forest ecologists, wildlife managers, and politicians on how best to manage public forests and the human/forest interface. This topic is ideal for students to examine from a variety of perspectives, since it involves everything from predicting fire behavior and protecting public health from air pollution, to managing wildlife and fisheries in over-logged or burned environments. It is likely that each university has similar topics of local interest that could be used to generate projects that would capture the interest and imagination of their students.

By the time students reach their junior and senior years, and are taking more specialized courses in their fields of interest, our Renaissance campus would require that they participate in at least one independent research project, ideally with other students from different disciplines. They also would be required to write up and present their research, either in a classroom setting or campus-wide undergraduate research symposium, or both. A number of avenues exist on most campuses to accommodate this kind of interdisciplinary research experience. For example, many campuses have introduced undergraduate research opportunities in which students work in a lab for a summer or a semester. This requirement would simply formalize what is often an otherwise informal learning experience. Another alternative would be to fit it within the existing "independent study" format, although we would prefer to rename it "interdisciplinary study" and would like to see at least two faculty members and at least four to six students from different departments participate in each one. However it is woven into the academic infrastructure, we believe that all students should have at least one research-like experience working on an open-ended, "real-world" problem before they graduate.

These are modest proposals that could readily work within the existing structure of most colleges and universities (and indeed are already happening at some institutions around the country). One of many examples that illustrate this approach is the Center for the Study of Institutions, Population, and Environmental Change (CIPAC) at Indiana University, whose mission is to understand how and why some forests are fragmented, degraded, and losing species, while other forests are in good condition and even regrowing and expanding. This long-term study (Dietz et al. 2003) involves environmental scientists, geographers, political scientists, satellite imaging experts, computer scientists, statisticians, and students at all educational levels.

For these programs to work and become more common, we will need to radically change the focus of the university as a faculty-centered research and teaching environment where students come to be taught, to one that is known for being student centered, where students participate in research and come to learn. Despite widespread belief to the contrary, there is a natural and important connection between teaching and research. Faculty engaged in creative research endeavors can bring the excitement of discovery and up-to-date scientific advances to the courses they teach. Active faculty researchers can model science as a human endeavor, involving diverse people in traditional and nontraditional career settings (AAAS, 1990). Moreover, they can model and foster rigorous critical thinking, originality, creativity, and problem solving with students at all levels of educational experience, as well as in diverse educational settings.

The Renaissance Campus would benefit all students, even those who do not go on to become members of biological research teams or biotech industrial groups. Graduates of such a campus will have developed skills that ensure that they can think critically, communicate with others, and be intellectually flexible in whatever career they do pursue. As important, they will have developed the skills they need as citizens to look at questions of local, national, and global concern, and make informed decisions that can potentially affect us all.

Acknowledgements. This manuscript arose from the conversations begun with participants at the "Meeting the Challenges" workshop in March 2003. We especially thank Diane Smith for helping us synthesize our ideas and develop this manuscript. Brewer's participation was funded by a grant to the University of Montana from the Howard Hughes Medical Institute.

References

American Association for the Advancement of Science, *Science for all Americans: Project 2061*, Oxford University Press, New York, 1990.

Brewer, C. A. and L. J. Gross, "Training ecologists to think with uncertainty in mind," *Ecology* 84 (2003) 1412–1414.

Clark, J. S., et al. "Ecological forecasts: an emerging imperative," *Science* 293 (2001) 657–660.

Dietz Thomas, Elinor Ostrom, and Paul C. Stern. "The struggle to govern the commons." *Science* 302 (12 December 2003) 1907–1912.

National Research Council, *Bio 2010: Transforming Undergraduate Education for Future Research Biologists*, National Academies Press, Washington, DC, 2003.

Thornton J. W. and R. DeSalle, "Gene family evolution and homology: genomics meets phylogenetics," *Annual Review Genomics and Human Genetics* 1 (2000) 41–73.

Bioinformatics and Genomics

Julius H. Jackson
Michigan State University

Julius Jackson is Professor of Microbiology at Michigan Sate University. At MSU Jackson directs a lab that focuses on genomic informational systems and dynamics, in particular on the information structure of genes, proteins, and prokaryotic genomes, and on gene organization and its role in prokaryotic physiology. Jackson received his PhD from the University of Kansas, and has held postdoctoral positions at Purdue University.

> The spectacular successes of the sciences is due in large measure to the use of mathematics in the creation and analysis of models for the phenomena of interest.
>
> — R. E. Mickens

Education across the disciplines that constitute bioinformatics poses a fundamental challenge for how to change the preparation of students and faculty to enable them to conduct research on contemporary biological questions. Much of the challenge resides in the special relation of bioinformatics to what is now called the "New Biology"—the data-rich biology of the twenty-first century. Weaknesses in the training of biologists who have good quantitative skills may result in missed research opportunities and slow progress in important areas of biology and medicine.

The first section of this report will explore the rise of bioinformatics to its current high-profile position at the center of the biological sciences. Subsequent sections will pose challenges and propose ways to implement strategies that may enable biologists to (a) train scientists across disciplines of mathematics, computer science and biology; (b) form multidisciplinary collaborations among scientists with cross-disciplinary training; and (c) educate students to learn across disciplines.

The Rise of Bioinformatics

Bioinformatics is a recently developed mainstream area that has become critically important to biological research. "Meeting the Challenges" in bioinformatics poses a special challenge, namely, to shape a working definition for bioinformatics that sufficiently includes the major fields contributing to the emergence of this new information science. As currently practiced, bioinformatics includes the development and application of computational methods to organize, access, interpret, and expand large static data sets of genomic sequences, macromolecular structures, and expression arrays. In practice, definitions vary according to the extent to which practitioners have interests in and understand the biology beyond algorithm development. Narrow definitions of bioinformatics lessen its importance as an emerging discipline and have often come to symbolize biologists' hiring computer scientists to make a tool that anyone can use without thinking. If this area becomes more comprehensively associated with the study of the structure and dynamics of information in biological systems, it is more likely to gain longevity and become established as a distinctive field of study.

Biology also crosses the quantitative skills boundary with other subjects, notably chemistry, biological and chemical engineering, and (in areas of public policy), economics. These disciplines deal extensively with a systemic view of biological entities, and it is possible to find examples in these areas that illustrate that it is possible to train students to work effectively at the interfaces of biology with other disciplines. However bioinformatics—the focus of this paper—is primarily about the joining of biology, mathematics, and computer science.

The integrative character of contemporary efforts to answer biological questions continues to expand the amounts and varieties of information stored in databases and, consequently, the need for methods to interpret and relate this data. Thus as the study of biology becomes more integrative, approaches to teach, learn, seek, and apply biological knowledge are inexorably evolving toward greater use of methods drawn from mathematics, statistics, and computer science. Breaking through new frontiers in biological research will require assemblies of cross-trained, multidisciplinary teams of scientists to engage in collaborative research that can model the function of biological systems in the real world.

Reductionist paradigm

Every living organism is a biological system, that is, a functional unit that receives input from its environment, processes that input, and elicits a response to its environment by producing some output. A standard historical practice of research in biology was to describe the features of organisms (more formally, phenotypes) in order to gather sufficient data to construct a conceptual model of the observed state of a biological system. Through much of the twentieth century, this research focused on observing and describing organisms at higher and higher magnification of smaller and smaller pieces. Not surprisingly, undergraduate curricula eventually followed research practices. Thus, even as biochemical (molecular) methods were being applied to study biological systems, the predominant practice in education remained to describe systems conceptually. Construction of descriptive, conceptual, static models was done for the purpose of explaining the function of biological systems.

Notable exceptions to this math-avoidance approach to biology were quantitative studies of enzyme reactions, energy transfer reactions, genetics, population biology and evolution. Those scientists practicing "quantitative biology" focused on the study of mechanisms. The introduction of chemistry, mathematics, and physics into biological research opened new frontiers, notably challenges to understand how biological systems worked. Exploration of those newly opened frontiers aimed to define the chemical and physical mechanisms that enabled biological systems to function, and to define function at the smallest possible scale. This "mechanistic" approach to biology was, necessarily, reductionist, that is, aiming to discover the smallest functional units and how they worked. Thus, for example, the study of the biology of plants and animals (including humans) moved from whole organism to organs; then to cells; then to subcellular organelles; then to biochemical pathways that power cells and make cell components such as macromolecules; and finally all of the way to the shifts of electrons and protons from one atom to another within a molecule.

Much the same approach was applied to the study of simpler organisms such as bacteria and other single-celled microorganisms. Results from these high-resolution reductionist approaches to biology produced steadily increasing amounts of data and incremental understanding of the function and structure of molecular components of individual cell types. However, while these subcellular studies provided highly refined insights into the structure and operational properties of cellular subsystems, it gradually became clear that knowledge of parts was insufficient to predict the behavior of the whole organism.

Individualized Research

Associated with the period of reductionist thought in biological research was the rise to supremacy of competitive individualism in the conduct of research. The peer-review process of distributing federal funding for research has been widely regarded as a funding mechanism that opened opportunities for research support to all competitors rather than making research awards only to those who were best known in a kind of "old boys club." Competition for funding has shaped tenure and promotion criteria. The influence of funding strategies that promote individual competition in research narrowed even more the specialization of focus and expertise for individual biological scientists. Collaborative research was structured to encourage scientists to retain their narrow specialties and to cooperate primarily with those within the same field who possessed specialized knowledge complementary to the collaborators'. In other words, collaboration was practiced to refine the reductive edges of disciplinary interfaces in order to achieve an increasingly higher resolution of the mechanistic focus that was required to describe how the smallest operational pieces of cells worked.

These practices generated substantial amounts of data about a few well-studied model organisms. The working assumptions were that a highly detailed knowledge about an intensively studied model organism, or model macromolecule such as a particular enzyme, would lead to better understanding of how the complete organism functioned and how organisms interacted with each other and their environments. The prevailing view was that in-depth knowledge about how biological systems worked required exhaustive studies in a select few model organisms that could illustrate how all organisms functioned. The study of model organisms very much remains a contemporary practice in biological research; examples of such models can be found on the website of the National Center for Biotechnology Information of the National Library of Medicine.[1]

Although this approach came at the expense of studies in evolutionarily diverse organisms, a great deal of success flowed from these highly focused, fiercely competitive research efforts. We learned that it is indeed possible to define a core set of metabolic properties that represents a similar set of properties in all organisms. Examples include the metabolism of glucose to produce energy, and the mechanisms of synthesis and breakdown of fats. The mechanistic approach, when applied to a broad variety of organisms, also revealed unique metabolic properties associated with the habitat of the organism. Perhaps most importantly, the genetic code and fundamental mechanism of DNA replication vary only

[1] www.ncbi.nlm.nih.gov/About/model/nonmammal.html.

slightly among phylogenetic groupings of organisms. Studies of mechanisms in model organisms did indeed produce knowledge of universal rules for certain biological functions and processes, including DNA structure, gene composition, chromosome replication, biosynthesis of macromolecules, and energy production and storage, to name only a few.

Integrative Biology

The impressive success of mechanistic studies of biological systems unintentionally contributed to the rise of interest in integrative biology. A standard practice in mechanistic studies of the physiology of organisms is to associate phenotypic properties under study with the gene or set of genes responsible for the phenotype. For example, what gene or set of genes is responsible for diabetes or cancer in humans? In bacteria, what gene or set of genes enables one bacterium to cause lethal disease while a close cousin is harmless? What changes in a gene can make an enzyme functional in boiling water for one prokaryote but unable to function above refrigerator temperature in another organism? The intense interest in the study of genetic mutations was greatly expanded by technological advances of gene cloning and sequencing. By creating a biochemical bypass to the requirement to develop a genetic system in every organism to be studied, these technologies opened many organisms to study. The explosion in production of genetic information thus began.

The push to understand mechanisms of physiology through the search for genetic mutants revealed that the function of systems studied often depended upon the sometimes subtle interactions of many genes involved in a broad, regulatory network. Therefore, even when mechanistic information was available for each component of a regulated metabolic system in an organism, it was not possible to understand the overall function of the system sufficiently to construct a working model using only the snapshots of information available from the mechanistic studies. Upsetting the reductionist paradigm, system function could not be inferred just by inspection of the mechanistic rules of each subsystem or system component.

Rapid techniques to clone and sequence genes opened the possibility to study any organism that could be cultured or from which DNA could be obtained. It therefore became suddenly possible to conduct mechanistic studies in any organism of choice, from thermophilic bacteria to humans. This technological revolution in gene cloning and sequencing, combined with the fervor to

Growth of GenBank

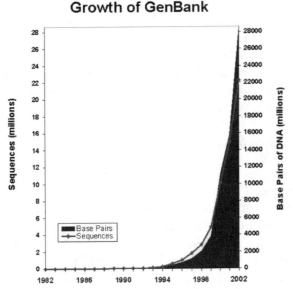

Figure 1. Growth of sequence data.

apply the new technologies to mechanistic studies, produced an exponential expansion of information on gene and genome sequences stored in newly constructed databases with free access to the public (see Figure 1).

Because of the vast potential of this new information on gene sequences to the emerging biotechnology industry, the newly established National Center for Biotechnology Information took on the responsibility of curator of the GenBank repository of nucleotide sequences of genes. Established in 1988 as a national resource for molecular biology information, NCBI creates public databases, conducts research in computational biology, develops software tools for analyzing genome data, and disseminates biomedical information—"all for the better understanding of molecular processes affecting human health and disease."[2]

These new activities of NCBI involved the joint efforts of mathematicians, statisticians, computer scientists and biologists. The new field of bioinformatics arose from the need to manage the information in multiple databases and make such information accessible to scientists in the public domain.

At the outset of bioinformatics, a problem of major concern was how to enable the mechanistic biological scientists to gain full access to the sequence data and help them learn how to search and use the data to support their individual studies. From the beginning, bioinformatics produced many computational tools for use in the study of information in nucleotide or protein sequence

---
[2] www.ncbi.nlm.nih.gov

databases. The unfortunate reality is that the masses of contemporary research biologists and students do not sufficiently understand how these bioinformatics tools work and consequently how to maximize use of the tools to support their research.

In some quarters an impression has evolved that computer scientists should simply develop software to support the biological scientists' research needs and make such software easily usable to produce results that are readily interpretable with little or no investment of intellectual capital. Although such software expectations may be desirable goals, reaching them will require the full collaboration of biologists, computer scientists, and mathematicians who all understand the problems to be solved and who fully collaborate in their solution. Examples of this kind of multidisciplinary collaboration have resulted in new software associated with whole genome sequencing projects and with the attempts to interpret expression array results. Unfortunately, too many scientists now learn to apply rapid chemical methods that stuff databases with large amounts of data that potentially harbor unintelligible information.

Multidisciplinary collaborations among cross-trained computer scientists, mathematicians and biologists are necessary to make sure that the stored information is interpretable. One goal emerging from application of bioinformatics tools is to manage these ever-expanding databases of information in ways that enable discovery of knowledge about system behavior at every scale. One of the greater expectations of the "genome revolution" is to be able to see every letter of the genetic code for any organism, to be able to mine sequence and other databases for information from which to infer both characteristics of physiology as well as potential responses to particular environments.

Biology in the New Century

Scientific questions about the operation and interaction of complex biological systems are setting the framework for extraordinary and exciting discoveries as biology becomes the premier science of the twenty-first century. In this century, the general approach to study biology will progress, from the traditional principle of *superposition* that dominated the twentieth century research, to incorporate more of the approaches of systems analysis. The principle of superposition refers to the assumption that "a complex process can be decomposed into constituent elements, each element can be studied individually and reassembled to understand the whole" (West,

1990). This principle describes the linear thinking that had so much success during the twentieth century in analyzing physical phenomena. It is the basis for the reductionist (mechanistic) approach to the study of biological processes and cell and organism function described above.

In contrast, the mainstream of biological research in the twenty-first century will be conducted by (a) treating biological processes as systems, (b) constructing quantitative models of biological systems, (c) converting qualitative models to quantitative models, and (d) using mathematics as the language of biology. Whereas qualitative models form the basis for speculation, quantitative models form the basis for simulation. Models used for simulation present opportunities to assess how accurately they describe a real-world system behavior and serve to guide design of critical experiments to test system properties. Some modeling packages (e.g., Stella,[3] VisSim[4]) allow fairly complex mathematics implementations with a minimum of specialized training. These packages provide an introduction to modeling and simulation for individuals who have little background in these areas. As mathematics is the language of science, so mathematics is becoming the language of biology (Figure 2).

In this new paradigm nearly every biological research study will need to use some aspect of bioinformatics in order to make reasonable progress. The reason for this is that most biological research will benefit from genomic sequence and expression data. The potential impact of the multiple and expansive biological databases on planning, implementation, and analysis of biological research is too great to leave the power of this information to any one disciplinary interest. Thus all scientists who will engage in or be expected to understand biological research will need basic knowledge of critical principles in mathematics, computer science and molecular biology.

Understanding the function of any biological system requires integration of rules discovered in the studies of mechanisms regulating structure and process. Successful biologists in this new century will need to adopt practices and strategies for research that will enable them to:

- develop a systems view of their research focus;
- construct quantitative models from qualitative descriptive models; and
- apply quantitative models to simulate behavior of biological systems.

---

[3] www.iseesystems.com/
[4] www.vissim.com/

A New Century Of Biology

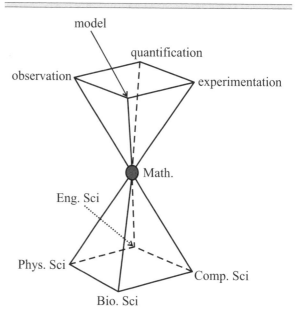

The Language of Science is the Language of Biology

Figure 2. Pyramids of model Building

Curriculum

As bioinformatics has emerged over the past decade as a new interdisciplinary field, computer scientists and mathematicians have joined with molecular biologists to produce tools required for data mining and exploration. Commonly, undergraduate curricula either treat bioinformatics as a new discipline or develop special tracks within existing majors that emphasize some subset of the broad field of bioinformatics. Workshop discussions revealed widely divergent views about what should constitute training in bioinformatics. More troublesome, perhaps, is a not uncommon sense of resentment within the communities of computer scientists and mathematicians when their fields are treated merely as service providers for biology programs.

A simple requirement for the development of bioinformatics tools is that they be easy to use and as easy to interpret as to use. Ironically, this quest for unambiguous interpretation is in part responsible for mathematical scientists' perception of using algorithm and database developers only as service providers. Too often, biologists use the tools of bioinformatics to study biological questions without adequate knowledge of the limitations of these tools. Methods are accumulating faster than biologists can learn how best to apply them to increase understanding. Appropriate education in mathematics and applications of mathematics to study the behavior of biological systems will accelerate new discovery in these ever-growing databases.

As we have argued, biology is a quantitative science based upon careful measurements to model complicated and complex natural systems. Yet the current curriculum for an undergraduate degree in many fields of the biological sciences provides weak quantitative preparation and exposure that limits options of the graduate for future studies and employment. Undergraduates interested in professional careers in medicine, dentistry and related areas comprise the vast majority of majors in the biological sciences. Additionally, "service courses" for many areas that depend on biology, (e.g., allied health; pharmacy; engineering; physics; and chemistry) constitute a substantial part of the teaching load for biology department faculty.

The quantitative expectations established for admission to professional schools in health-related fields are substantially lower than the capability of the students and the lowest in all of the natural science. It is likely that this low expectation drives the undergraduate biology curriculum to exclude quantitative rigor in favor of qualitative description and memorization in the major courses. Typically, students flock into biology programs to escape the mathematics associated with the other natural sciences and engineering. And students interested in careers in scientific research and education in the biological sciences have a curricular exposure to mathematics indistinguishable from pre-professional students.

Thus, the low quantitative expectation set for admission to professional schools negatively influences the preparation of students planning research and teaching careers in biology. We are now at a point where these low expectations and poor quantitative preparation impede novel research discovery at a time when databases with all types of biological information are increasing exponentially (Figure 1).

The undergraduate curriculum must change to raise expectations and meet the challenge to educate a new biological scientist capable of joining in research collaboration with mathematicians, computer scientists, physical scientists, and engineers to solve biological questions.

Curriculum Reform

A curricular awakening to greater inclusion of mathematics for basic biological education has been quietly underway across the country. The broad sense of need for bioinformatics has contributed substantially to this trend. The academic response to perceptions of greater need for bioinformatics has been to develop new courses and in some cases to develop bioinformatics as a new major.

Beyond bioinformatics, *Bio 2010* (NRC, 2003) documents the increasing importance of changing the fundamental approach to educating biologists for this new century. *Bio 2010* poses the challenge of how to develop scientists who are *interdisciplinary*—the term most often used to describe the new training paradigm.

The term "interdisciplinary" is related to "multi-disciplinary" and "cross-disciplinary," and often these are used interchangeably. For the purposes of this chapter, however, it will be useful to consider some subtle distinctions among these terms, especially in regard to how these terms apply to meeting the challenges in bioinformatics.

- *Multi-disciplinary* education or training is analogous to pursuing multiple disciplinary majors, e.g., simultaneous majors in microbiology, mathematics, and computer science, or a double major in mathematics and molecular genetics; or computer science and biology. This approach is one that some particularly talented students choose. It does not benefit from mutual reinforcement in multiple courses, but rather it is a pursuit of the standard understanding of the standard examples for each discipline. Little effort goes into reaching across the boundaries of mathematics for biological examples to illustrate a mathematical application, for example.

- *Cross-disciplinary* education is analogous to pursuing the combination of a major and one or two minors, e.g., microbiology major and mathematics minor, or any combination of biology, mathematics and computer science.

- *Interdisciplinary* education is pursuit of several traditional disciplines, each with less than a standard major curriculum. It is more like combining coursework for a minor in multiple areas with a major in none of them. Thus an interdisciplinary education borrows from multiple disciplines to form a particular thrust that can define or establish a new discipline.

Bioinformatics reflects the interdisciplinary concept of forming a new field of study. The emerging pivotal role of bioinformatics in the conduct of biological research leads to the question of how to effectively train the "new biologist" who is capable of exploring these new frontiers. Ideally, colleges and universities should strive to implement multidisciplinary educational collaboration by professors to teach major and minor courses in a cross-disciplinary curriculum so that the graduate develops a specific area of expertise combined with the necessary knowledge to communicate effectively with other disciplines. This approach to education leads to the desired goal of multidisciplinary team collaboration of scientists with cross disciplinary training.

What course content would adequately prepare biologists to understand how best to apply quantitative methods as well as to collaborate with mathematicians and computer scientists? A well-prepared undergraduate with a major in a biological science and a near-major in mathematics with several computer science courses would have the technical background to speak across multiple disciplines and could bring adequate foundation knowledge for cross-disciplinary collaboration. Such preparation would include, for example, calculus through ordinary differential equations, descriptive and non-parametric statistics, linear algebra, real analysis, stochastic processes and probability, nonlinear mathematics, discrete mathematics, systems analysis, and programming in PERL, C++, and MATLAB.

Yet just requiring biology students to take more courses in mathematics and statistics would not accomplish the goal of preparing large numbers of graduates to collaborate across disciplines. This realization was repeatedly emphasized during "Meeting the Challenges" workshop sessions. Instead, workshop participants recommended five inter-related steps to meet the goals of developing the desirable cross-disciplinary interactions:

1. *Make the core course requirements for mathematics and statistics for biology majors the same as requirements for chemistry majors.* Biology curricula are now heavily based upon substantial course exposure in chemistry, i.e., equivalent to a minor, but without at least the mathematics preparation required for chemists. Typical current mathematics requirements for biology majors are either calculus I and descriptive statistics or calculus I & II. New minimal recommendations are: calculus I and II (two semesters), ordinary differential equations (one semester), descriptive statistics (one semester), and non-parametric statistics (one semester). (Although in some programs a third semester of calculus is required to take ordinary differential equations, this is not always the case.)

2. *Develop a first course for biology majors that includes or features mathematical applications.* Mathematics and computer science majors would join the same course with biology majors. One way to achieve this would be to focus the first semester on integrative biological and ecological systems and the second semester on cellular & molecular biology. The goal should be to raise the exposure of biology students to quantitative perspectives.

3. *Wherever possible, introduce substantially more mathematics into existing biology courses.* This recommendations may be the hardest to accomplish because it requires faculty assigned to key courses to embrace the idea of increasing quantitative expectations. Many faculty will resist this approach. Realistically, this may be accomplished for only a few courses.

4. *Develop one or two courses on mathematical methods of modern biology.* One approach is to focus a sophomore-level course on Concepts of Modeling and Analysis of Biological Systems. Calculus I would be a prerequisite. Another strategy is to create a junior or senior course on Mathematical Principles for Bioinformatics. Offer this course in a curricular context that involves students from the mathematical, computer, and biological sciences.

5. *Offer a computer-based course as an applied laboratory in bioinformatics.* The course would utilize existing algorithms, computational tools, and databases to analyze genomic or other "...omic" data. This type of course can be aimed in many different directions, for example, human genetics and genomics, microbial genomics and ecology, forensic science, or pharmaceutical drug design, to name just a few.

These changes can present substantial challenges to small and large departments alike. Many approaches and strategies are possible. Some programs employ strategies that expose all biology students to bioinformatics, while others treat bioinformatics as a separate, competing discipline. Some examples:

- The University of Maryland integrates bioinformatics across the course sequence for biology majors.
- The University of Wisconsin-Parkside offers a major in bioinformatics.
- Rochester Institute of Technology offers a new undergraduate bioinformatics degree program through its Center for Biotechnology Education and Training. The program involves 17 new courses.
- Foothills College offers a certificate in bioinformatics through its biotechnology program.
- Capital University offers a comprehensive major in computational science including bioinformatics.
- Wheaton College (MA) links computer science and biology courses to teach bioinformatics.

BEDROCK Faculty Development

At Beloit College faculty are helped in their efforts to incorporate bioinformatics and quantitative methods into biology curricula through a faculty development program called BEDROCK: Bioinformatics Education Dissemination: Reaching Out, Connecting and Knitting-together. BEDROCK aims to develop and support classroom resources for bioinformatics education in the realization that "bioinformatics is central to biology in the 21st century." The project consists of 18 faculty workshops in bioinformatics problem solving.

BEDROCK grew out of the BioQUEST Curriculum Consortium led by John Jungck at Beloit College. BioQUEST's focus is to educate biology educators to see and teach biology as a quantitative science. One strategy focuses on the three P's: Problem Posing; Problem Solving; and Peer Persuasion.

Web Sites:
BEDROCK: www.bioquest.org/bedrock/
BioQuest: www.bioquest.org
The 3 P's: www.bioquest.org/index3ps.html

Other examples can be found elsewhere in this volume, and on the internet.[5]

Associated with each curricular reform is a parallel challenge of faculty development. One approach is described in the sidebar on BEDROCK, a project at Beloit College aimed at raising faculty expertise in bioinformatics. Others are the workshops on computational biology sponsored by the National Computational Science Institute (NCSI).[6] More efforts like these would go a long way to "meet the challenge."

The emphases on bioinformatics in education and research appear to constitute the major force for change to instill greater quantitative awareness and facility in biology education. While this thrust is much needed and should be considered a welcome influence on biology, curriculum revision should aim to enable biology, mathematics and computer science majors to think and work in integrative concepts of biological systems and how they work. Thus it is especially important to prepare biology students, mathematically, for research beyond applications of bioinformatics.

While some efforts at undergraduate curriculum development and revision are supported by external resources, most examples represent local faculty efforts to change curriculum without external funding. The current budgetary crises in which most colleges and universities operate present special challenges for faculty to find time to revise curricula, especially to shift major training and educational paradigms. Such paradigm shifts need strong support

from traditional sources for curriculum revision. The National Science Foundation, Division of Undergraduate Education, has long been a source of support for curriculum development. A recent "Dear Colleague" letter directly addresses support to supplement existing NSF grants for the changes discussed here. In particular, NSF has announced a special program in Interdisciplinary Training for Undergraduates in Biological and Mathematical Sciences.[7]

At the graduate level, in 1999 the National Institutes of Health (NIH) initiated a predoctoral training program to produce doctorates in the then new discipline of bioinformatics and computational biology. It hoped to identify and recruit applications from "students with strong quantitative skills from biological backgrounds and students from computational backgrounds with an interest in biological problem solving." The NIH initiative provided no direct support to modify undergraduate education but it did point to the need for larger pools of biologists with strong quantitative skills, and of mathematical sciences graduates with interest in solving biological questions.

Faculty

An extensive research training regimen for biological scientists has evolved to include at least two or more years of postdoctoral research studies and often more than one postdoctoral research experience. All of this comes before a biological scientist obtains a "permanent" position in the tenure stream of an academic institution or a research position in industry.

[5] www.cbs.dtu.dk/dtucourse/specialization.html

[6] www.computationalscience.org/workshops/summer04/index.html#biology

[7] www.nsf.gov/pubs/2003/nsf03037/nsf03037.htm

The dilemma of too many postdoctoral researchers for the available number of permanent positions in the U. S. has long promoted a training atmosphere in which the emphasis on numbers of publications limited the researcher to a narrow disciplinary focus in an effort to maximize productivity. University faculty strive to gain and sustain research productivity by publication of peer-reviewed research articles and books. The intensity of research focus required for competitive productivity in a discipline tends to discourage faculty pursuit of novelty in direction or method. It should not be surprising, then, that a major challenge for the type of curriculum revision proposed here is how to encourage and reward faculty to participate and contribute to major changes.

Faculty who teach undergraduates tend to teach a lot, and teaching loads will either support or discourage faculty contributions. Institutional support to reduce faculty teaching loads is essential to implement strategies to include more mathematics in biological sciences curricula. Institutional support comes from many levels and can take many forms. In order to succeed, creative strategies developed at a departmental level need the support of deans and higher-ranking administrators. Department-level strategies for faculty support may include:

- Team teaching. (In medical schools, team-teaching strategies routinely work to *reduce* course teaching loads, and could be more extensively used to increase research time available for faculty teaching undergraduates.)
- Revision of teaching load accounting. (Count teaching load by course participation rather than contact hours.)
- Research and teaching collaborations. (Encourage intramural faculty collaboration among biologists, computer scientists, and mathematicians in both research and teaching.)
- Weigh more heavily faculty teaching contributions to multidisciplinary courses.
- Offer traditional service courses in larger class sizes. (Perhaps teach pre-professional school biology as a service course different from biology majors. This would give faculty the freedom to change courses for biology majors to reflect leading edge combinations of mathematics and physical sciences to create more quantitative bases for modern biology.)
- Let graduate students or instructors teach service courses.
- Make calculus a prerequisite for core courses in the curriculum for a major in biological sciences.

- Encourage experiments with new or non-standard strategies for teaching and learning.
- Offer salary, travel and lodging support for faculty retraining activities.

Constructive change requires effective leadership at departmental and program levels, but also at the level of deans and higher. The first challenge—often surprisingly difficult—is to gain strong support from an entire department. Change is never easy in academia; as has been noted, marrying the cultures of biology and mathematics is especially difficult. When faculty do agree, academic officers need to provide budgets sufficient to enable department chairs to implement constructive change. Moreover, it is likely that some of these suggestions may run contrary to current tenure and promotion policy at many colleges and universities. Successful change in curriculum and faculty preparation to implement new curricula will require not only greater investment of resources by deans and provosts but also greater cooperation between administration and faculty to support the new directions in biology education.

Academia alone cannot shoulder the cost of curricular reform and associated faculty retraining. Federal agencies that influence public policy must boost support for curriculum revision by providing funds for curriculum change and faculty development. The National Science Foundation has long been the lead federal agency supporting postsecondary education of scientists in all scientific disciplinary areas that are not directly involved with health care. The comprehensive challenge of attracting new, quantitatively oriented students into the biological sciences and to interest traditional students of biology to pursue quantitative directions will best benefit from a combined effort of multiple federal agencies. The interests of administrators of NIH and NSF programs in supporting the "Meeting the Challenges" workshop points positively toward multi-agency support of the principles discussed here.

Most federal granting agencies and private foundations would readily support more experiments for change if they had more funds. Lacking new money, however, it is prudent to think about strategies that may be implemented without infusion of major new funds. A few examples:

- *Early career research development awards.* The long-standing support by NIH for the early career development of scientists in various, traditional research disciplines could substantively encourage young scientists to pursue new career directions in cross-disciplinary collaborations as quantitative biologists.

- *Mid-career research development awards.* A mid-career shift in disciplinary research direction or a change in the principal field of research can be terrifying to faculty, and disrupt the all-important research productivity. Special mid-career awards designed to help cushion the risk of change would help many scientists—biologists, mathematicians, computer scientists—explore the new cross-disciplinary fields of quantitative biology.
- *Workshops on change for university administrators.* Such workshops (similar to many now designed for new department chairs) would help academic administrators learn to provide positive support for change in the research and education paradigms without antagonizing faculty or creating negative side-effects.

An especially promising observation is that the cooperative activities associated with "Meeting the Challenges" involve several professional organizations of scientists and mathematicians working in concert with federal government agencies along with academic faculty and industrial researchers. This type and degree of cooperation bode well for the likelihood that sufficient resources may emerge to support key strategies and programs needed to implement the new educational and research directions.

Students

The most important requirement of "Meeting the Challenges" is for undergraduate students in each of three distinct disciplines—a biological science, mathematics, and computer science—to gain insight and understanding on how to apply the other two disciplines to define and analyze biological questions. The typical current preparation in each discipline is insufficient to prepare a student for participation as a full partner in information age biology of the twenty-first century.

For prospective biologists, the first step on this new path is to prepare to study mathematics as an integral part of biology research and teaching. For mathematics students, the first step is to prepare to study biology as sets of complex systems awaiting mathematical analyses. For computer scientists, the first step is to look beyond programming to frame biological questions that algorithm development can help answer. Students and practicing scientists in each area have long-standing limitations or irritations to overcome.

- Computer scientists often feel relegated to programmer status instead of being full participants in a collaborative study using bioinformatics. Limited knowledge of mathematics and biology may confine the computer scientist to the role of tool development.
- Many pure mathematicians avoid applications of the mathematics they develop, while applied mathematicians shop for an application of a method they developed and prefer to use. Each could help by learning sufficient biology to understand key questions and define problems that may yield to mathematical solutions.
- Typical biological scientists are refugees from mathematics, and rely on descriptive qualitative models to describe biological systems. They tend to "throw experiments at a problem," but limit research designs to their zones of comfort and familiarity.

Each such problem, admittedly a bit of caricature, would yield to a collaborative team solution.

The serious undergraduate student with an interest in a career in biological research or education can approach that interest from many directions and still be well prepared as long as that preparation includes an understanding of how to work in teams to solve problems. The team approach is no substitute for individual responsibility: it must be accompanied by a commitment to master subject matter and self-motivation to pursue educational goals. However, any undergraduate who might pursue a research career should expect to engage in an educational collaboration with the faculty to achieve specific goals.

One of the more important student-oriented changes that needs to occur is a change in patterns of recruiting and advising. Many of today's quantitative biologists were not originally trained as biologists. Part of the reason, as we have noted, is that the biology curriculum is not conducive to quantitatively oriented courses. But another reason is the tradition of biology students escaping mathematics-intensive courses. Along with curricular changes, we need to encourage mathematically proficient students to look toward biology as an outlet for their skills.

Students intending to pursue graduate studies and research careers should take courses designed for the needs of majors, not the less rigorous service needs of curricula leading to professional school and professional degrees. Moreover, undergraduate students majoring in mathematical, computational, biological and physical sciences should share time in disciplinary courses designed for the major. The inclusion of quantitative biological models in such courses will reinforce the connections of all fields to biological questions.

Some argue that since many pre-professional students actually wind up in research careers, it is a mistake to encourage any two-track system where the large number of pre-professional students receive a less rigorous foundation. Under this system, those who change career tracks—including many very good students—might have to retake courses to make up the rigorous material that they missed. Thus, in single track programs, this track should be as quantitatively rigorous as possible; in dual track programs, strong advising will be necessary to ensure that all capable students receive the more challenging preparation.

Undergraduate Research

Notwithstanding all other curricular planning and educational efforts, the most reliable indicator for scientific career choices are experiences with research as an undergraduate. Exposure to "real science" during college (or earlier) is a proven strategy to simulating interest in scientific careers. To stimulate interest in the quantitative "new biology," students majoring in mathematics, computer science and biological sciences should participate in cross-disciplinary undergraduate research activities.

Support for undergraduate research is provided by the NSF in the form of Research Experience for Undergraduates (REU) supplements to existing research grants and REU site program grants that have a thematic basis for the award. Similar types of support for undergraduate research is available from the NIH. While these research experience opportunities have been oriented toward disciplines, cross-disciplinary exposure is gaining increasing emphasis. The REU sites program at NSF has begun to move in this direction by promoting the entry of mathematics and physical sciences majors into biological research.

Multidisciplinary Teams

The emphasis on multi-disciplinary team solutions to biological questions must be instilled in undergraduates from the outset of their educational exposure. Whenever possible, student participants in the undergraduate research experience should be offered a research opportunity that involves multi-disciplinary faculty collaboration on a biological problem. A multidisciplinary team approach, involving mathematicians and biologists, has worked well at Michigan State University to provide cross-disciplinary research experiences for both undergraduates and high school students.

A variation of this strategy can be seen in the BioQUEST Curriculum Consortium's emphasis on the "Three P's": Problem Posing, Problem Solving, and Peer Persuasion. These illustrate a "best practice" for introducing undergraduates to research: add the three P's to multidisciplinary team exposure as a recipe to prepare undergraduates for "Information Age" studies in biological sciences research and education.

Finally, in thinking about how to "meet the challenge" of providing faculty who can teach quantitative biology, we need to keep in mind that nearly 40% of PhD degrees awarded in science and engineering in the U. S. go to foreign students, many of whom are supported by federal research grants to scientists at American universities. U.S.-born students seem to be less aware of the opportunities to receive graduate education completely supported by federal research dollars. While women continue to approach parity in the U. S. scientific workforce, except in mathematics and computer science, historically excluded and underrepresented minorities do not.

Programs designed to address issues of underrepresentation, such as those in the Division of Minority Opportunities in Research (NIGMS) at NIH, are also emphasizing the early introduction of undergraduates to research experiences. Although still small, the pool of black, Hispanic and American Indian students receiving baccalaureate degrees in science and engineering is gradually increasing. These underrepresented groups, including women in mathematics and computer science, represent a rich resource of U.S. scientific talent that remains largely underutilized. Strategies to revise how we train biologists, computer scientists and mathematicians to enable them to work in multidisciplinary teams must especially apply to these historically excluded and underrepresented U. S. citizens. The notable success of Freeman Hrabowski's Meyerhoff Scholarship Program[8] for undergraduates at the University of Maryland, Baltimore County, belies anecdotal assertions that a paucity in numbers of "qualified minorities" makes targeted graduate school recruiting efforts not worthwhile.

Providing the opportunities for early exposure of all undergraduate students to challenging research experiences must be a core thrust in the effort to prepare new scientists to exploit new information for the new biology.

References

Bruce J. West, "The Disproportionate Response," In *Mathematics in Science*, R. E. Mickens (ed.), World Scientific, 1990.

National Research Council, *Bio 2010: Transforming Undergraduate Education for Future Research Biologists*, National Academies Press, Washington, DC, 2003.

---

[8] www.umbc.edu/Programs/Meyerhoff/

Adapting Mathematics to the New Biology

Leah Edelstein-Keshet
University of British Columbia

This chapter addresses issues confronting the education of biologists in mathematics, quantitative methods, and their practical applications. Until recent decades, biology had been a descriptive science, with little need for a mathematical basis. (Exceptions were notable in certain parts of ecology and physiology, where a tradition of modeling is longstanding.) The development of the genome project, and then of post-genomic, data-intensive facets of biology have led to a profound need for sophisticated new techniques, including mathematical and computational ones. However, the expertise needed to push these developments forward is lacking in the biological community.

Until recent times, a biologist could function well with little or no mathematical training. This is no longer the case. One of the main difficulties is that the pace of innovation in biology has far exceeded the pace of renewal in training programs for future practitioners. This situation requires urgent attention if we are to educate a competent workforce for the current and future needs of biology, biotechnology, and the post-genomic life sciences.

In this paper we discuss the current status of curricula and suggest how to overcome shortfalls. We then survey the perspectives of faculty and students who are facing the need for change and address how such change could best be accomplished. Some examples of promising directions are highlighted in sidebars.

Curriculum

Current undergraduate curricula in mathematics and biology lag behind the needs of future life scientists who will face the challenges of modern biology. Widespread mathematical illiteracy is compounded throughout many levels of education: mathematics is poorly taught in school and college, and is consequently not understood, learned, or retained. This affects the market for educational material in many disciplines. Biology texts that are devoid of formulae and equations are more accessible, and hence more marketable. This has meant that, over time, the prevalent biological course materials are those that omit mathematics, even in instances where mathematics has an essential role to play. Obvious examples of population growth, neural conduction, or other physiological case-studies involving fluxes, mass balance, etc, are often handled with diagrams and "word equations," rather than common mathematical notation. Such examples mislead students into thinking that mathematics has no role in their biological learning.

A former president of the Society for Mathematical Biology, Leah Edelstein-Keshet is Professor of Mathematics at the University of British Columbia in Vancouver. Keshet is the author of *Mathematical Models in Biology* (Random House, 1988), recently reprinted by the Society for Industrial and Applied Mathemaics (SIAM). Keshet received her PhD in 1982 from the Weizmann Institute of Science, Rehovot, Israel. Her research interests include simulations of neuroinflammation believed to be associated with Alzheimer's disease, the dynamics of actin filaments and cell motility, and swarming behaviour (e.g., aggregation, migration) observed in a variety of social organisms such as insects, fish, and fowl. *E-mail*: keshet@math.ubc.ca

Mathematics 102/103
University of British Columbia

The Mathematics 102/103 sequence at the University of British Columbia contains many elements common to most first year calculus courses (derivatives, integrals, rates, areas, etc). As such it is a non-terminal sequence: students can go on to take later mathematics courses that require a calculus prerequisite. However, in this "biologically flavored" calculus sequence, many applications have been geared towards life science students. For example:

- Graphs and properties of power functions and polynomials are introduced in the first few lectures. A discussion about the rate of increase of volume versus surface area of a growing spherical cell leads directly to conclusions about how big a cell can be: balance of nutrient uptake (proportional to surface area) with nutrient metabolism (proportional to volume) imply that the cell cannot continue to expand beyond some size. This example introduces the idea of modeling, in a context with simple geometry, clear statements of assumptions, and very straightforward algebra. Moreover, the conclusions are of fundamental importance. They lead to further exploration of ways in which cells meet the challenge of diffusion limitation (e.g., by growing as thin cylinders or by becoming flat or ramified).

- Optimality principles abound in biology. The examples used vary from year to year, but often include optimal foraging. Here simple differentiation and finding optima can be used to predict how long an animal should stay in a given food-patch in order to maximize its net energy gain.

- Traditional calculus problems based on displacement and velocity of cars, planes, or trucks, can be easily replaced with biologically interesting examples of calculations involving moving particles or cells. Students delight at descriptions of the intracellular parasite *listeria* and its erratic motion inside a cell. Graphs taken directly from biological papers about the displacement (and velocity) of such tracked particles are useful examples of direct relevance to cutting edge research in molecular biology. Such examples motivate the idea that understanding velocity and rates of change are just as important in modern biology as in physics or in other sciences.

- The binding of receptors and ligands and their simple mass-action kinetics can be easily described at an elementary level. In the first weeks of a calculus course, this setting provides a strong example of modeling at work, even before rates of change and derivatives are defined. One can under-

An appreciation of mathematics cannot be built overnight. Inherently, mathematics is a subject built on many layers. A firm foundation in elementary school is the basis on which high-school mathematics stands. Similarly, solid high-school training is essential for college-level courses. However, most students are unaware of this layered construction until, too late, they flounder at the university level and then either drop mathematics or fail. Those with weak mathematical backgrounds often look to the life sciences as an escape hatch where this weakness was (traditionally) not a disadvantage. This is no longer the case for many of the promising avenues of biological research and biomedical practice.

In what follows, we focus primarily on potential changes in mathematics. We first consider the steps that could be taken to make mathematics more appealing, and hence more relevant to other disciplines, particularly to biology.

Mathematics courses and programs

One way to increase learning and retention of mathematics is to make undergraduate mathematics courses more interesting and attractive to students of all backgrounds, including those specifically intent on biology as a major. A number of strategies could be considered:

1. Instructors can illustrate the usefulness of mathematics as a tool by increasing the number of examples and problems that have biological relevance. Short modules can be used to highlight simple mathematical concepts in relation to interesting biological applications. (See sidebar above for some typical examples).

2. To make examples less artificial, instructors can revisit one or two biological themes in increasing levels of detail throughout a semester. At first, some simple algebra or graphical ideas could be applied. As students gain more technical expertise and learn new

take the formulation of balance statements that define steady-state fraction of bound C_1 and free C_0 ligand, and show, with suitable algebraic reduction and assumptions, that plots of C_1/C_0 versus C_1 fall along a straight line called a Scatchard plot (see www.curvefit.com/scatchard_plots.htm). Kinetic parameters that govern affinity of the ligand to its receptor can be identified from the slope and intercept of this line. A good opportunity for using real data occurs within this setting. Later in the course, once students have had exposure to derivatives and rates of change, the example could be discussed again from the perspective of differential equations.

- Population dynamics provides plentiful ideas that are fundamentally mathematical. Exponential and logistic growth are obvious examples of simple differential equations that can be studied analytically or geometrically (using direction fields or plots of dy/dt versus y). Adding a harvesting term immediately leads to the notion of population collapse under over-utilization, with ramifications towards ecology and resource management. (This is also one of the first simple examples of a bifurcation that can be described geometrically and intuitively.)

- Once students understand the predictions of a logistic differential equation, they can be introduced to the simple SIS epidemic model (so named to reflect the susceptible-infected-susceptible pattern of diseases in which no immunity is built up). Here the number S of susceptible and number I of infected individuals sum to a constant population size ($S + I = N$). This model reduces to a form identical to a logistic model, and reveals the surprising threshold behavior: If the population is small, the disease will be eliminated after a few people are infected, but if the population is large enough, endemic infection will result.

Many other examples of these sorts can be introduced at the first year level. Other modules include the visual response of a prey to an approaching predator (based on trigonometric functions and related rates), and the spacing of primordial teeth on the alligator's jaw (based on arclength and properties of a parabola).

In Math 102/103 the standard three hours of lectures are supplemented by simple computer labs based on spreadsheet manipulations. The purpose of these labs is to expose students to additional perspectives, including computational and conceptual ideas, related to the course content.

Further information: ugrad.math.ubc.ca/coursedoc/math102/
ugrad.math.ubc.ca/coursedoc/math103/

mathematical tools, the instructor can revisit topics and illustrate the use of more advanced techniques.

3 Real data from biology (and other disciplines) can motivate the need for analysis and quantitative thinking. Students might be encouraged to bring in a data set (from biology labs or courses, or by contacting biology faculty) as part of this exercise. This would encourage students to make a connection between what they learn in biology and the tools they gain in mathematics. (It would also, incidentally, raise awareness among biology professors that such links exist.)

These strategies are relatively easy for individual instructors to implement without extensive changes in course design or syllabus. They can also be used more extensively as the underlying philosophy for new courses. For example, a sequence of first year undergraduate courses developed at the University of British Columbia for students in the life-sciences stream exhibits some of these elements (sidebar above). Other courses at San

Diego State University (sidebar on p. 22) and Appalachian State University (sidebar on p. 23), to name a few, are even more closely built on a core of biology. While essential calculus topics are maintained, applications to biology are developed as illustrative examples and are used to demonstrate the usefulness of mathematical ideas in biological settings.

Although designed to capture the interest of biology students in mathematics, these steps also help attract talented mathematical students to potential careers in biology. Other ways of increasing interest include efforts that link colleagues in different departments or disciplines. These include:

4. Bringing in biologists as guest lecturers to motivate student interest in some research-related area that can benefit from mathematics.

5. Team-teaching a course or interdisciplinary program; this presents a greater challenge, as it requires formal approval from departments and faculties. It is also

Mathematics 121, 122
San Diego State University

Mathematics 121 and 122 is a sequence of two 3-hour first year courses at San Diego State University (SDSU) designed specifically for biology majors. SDSU is a large public university with students ranging widely in ability. The massive Math 121/122 sequence, under development by SDSU mathematician Joseph Mahaffy over the past 6-10 years, serves as a requirement for students declaring a biology major. Each term approximately 300 students enroll in this sequence (in sections of roughly 80 students each). The course consists of two hours of lectures and two hours of lab per week.

The course focuses on modeling, using realistic data as far as possible, to motivate learning mathematics and conceptual thinking. Although originally based on one or another calculus text, source material for the course was eventually developed from scratch by Mahaffy, and now consists of an extensive web-based set of lecture notes, labs, and homework problems, freely available on the author's website at SDSU. Weekly labs expose the students to Maple and Excel software in a variety of contexts that focus on novel applications to biology. A cohort of 30-40 students (working in pairs) learns about least squares fit of data and how to model trends or relationships mathematically. Lectures, labs, and homework are linked on common themes, including allometric, physiological, and ecological models, and methods of data analysis.

Further information: www-rohan.sdsu.edu/~jmahaffy/courses/s04/math121/index.html
www-rohan.sdsu.edu/~jmahaffy/courses/s04/math122/index.html

more costly and requires a greater degree of continual coordination. (Science One, discussed in the sidebar on p. 68, is an example of this approach.)

6. Linking a mathematics course to a biology course. This could be done in many ways, ranging from informal meetings of mathematics and biology faculty teaching parallel courses to agreements to synchronize one or two topics during the term. (The top sidebar on p. 69 offers one realization of this idea.)

7. Revising the undergraduate mathematics and biology curricula to match the new needs of the modern life sciences. This is much harder, as it requires departmental, faculty, and university approval, and impacts prerequisites, programs, and degree requirements. It may be even more difficult than these bureaucratic hurdles suggest, as the identification of the optimal curriculum is still quite controversial.

8. Introduction of a third-year mathematics course with minimal prerequisites on "hot" topics of computational biology: Is a DNA fingerprint sufficient evidence to convict an accused person of murder? How was the cystic fibrosis gene located? Or, more recently, how did a 40-year study of an Iowa family by scientists at Mayo Clinic locate a gene for Parkinson's disease? How was the mammalian DNA-based phylogenetic tree assembled, and how does it differ from the previous morphologically-based tree?

Core mathematical concepts required for undergraduate biology

The "new biology" is largely a biology based on data, models, and mathematics. Thus, to understand modern biology, students need increasing breadth of knowledge in mathematics, and deepening sophistication in its use.

All first-year undergraduate biology students need to master certain essential concepts from mathematics, including how to model simple biological systems; how to read and understand graphical information (e.g., scatter plots, histograms, pie charts); how to use units, dimensions, and scaling; and how to sketch simple curves. In addition, familiarity with several more advanced topics are now also essential:

- Rates of change versus total change (derivatives versus net differences).
- Unlimited growth (exponentials; log plots).
- Density dependence and saturating growth.
- Periodic behavior (phase, frequency, and amplitude of periodic functions).
- Linear versus nonlinear behavior (e.g. Malthus vs. logistic growth).
- Dynamic behavior, steady states, and stability.
- Discrete and continuous systems.

Ideally, these mathematical skills and concepts would be reinforced by suitable examples within biology courses in the early undergraduate years.

Math 1110/1120 for Biologists
Appalachian State University

Appalachian State University is a selective public university in Boone, NC with approximately 15,000 undergraduates. Math 1110/ 1120 for Biologists is an ongoing project led by mathematician Eric Marland and supported by a grant from the National Science Foundation. This sequence is taught to about 35 biology majors yearly. Small-scale initiatives such as this can act as nurseries for the development of material that can then be adapted to larger institutions.

The four contact-hours per week are held in a computer classroom where students work at individual terminals using software such as the spreadsheet *Excel* and the modeling and differential equation solver *Berkeley Madonna*. The material is based partly on texts by Wade Ellis *et al.* and Frederick Adler. Guest speakers from biology are featured several times each semester to enhance motivation and provide real examples of the usefulness of mathematics (for example, in drug doses).

Further information: www1.appstate.edu/~marland/classes/biocalc/biocalc.htm

References
Berkeley Madonna: www.berkeleymadonna.com/
Adler, Frederick R., *Modeling the Dynamics of Life: Calculus and Probability for Life Scientists*, Brooks Cole, 1998.
Ellis, Wade, et al., *Calculus: Mathematics and Modeling*, Pearson Addison Wesley, 1997.

Further, as students progress in their mathematical and scientific studies they should be exposed to the distinction between stochastic and deterministic systems. A number of topics from probability and statistics are especially important, including probability distributions; mean, median, and variance; how to draw inferences and to extrapolate from data to population characteristics; confidence intervals and their relation to precision of inferences, and experimental design (how to plan experiments so as to obtain optimal data).

At the second and third year level, elements of Markov Chain theory (now used heavily in biology) should be included. Important advances in Monte Carlo simulation have made Bayesian statistics accessible to scientists. These methods are especially useful for parameter estimation. Bayesian inference techniques are dominant in genetic analysis software, so biologists need substantive exposure to these techniques.

Challenges

For at least two major reasons, this "wish list" of mathematical prerequisites for biology poses immense curricular challenges. First, mathematics courses are expected to convey the coherence of (some part of) mathematics, not just isolated topics that happen to be useful in another subject such as biology (or business). Second, mathematics courses rarely are streamed so efficiently as to include only students of one single major—even one as large as the biological sciences. Both students and math-ematicians expect standard undergraduate courses to cover a relatively well-understood canon of core material that is not at all the same as the outline of topics presented above.

Some faculty have responded to this challenge by designing a coherent first-year mathematics course exclusively for biologists. Web sites for two such examples of bio-calculus courses designed by Joe Mahaffy and Eric Marland are listed in the sidebar on p. 25. Others compromise with a course that meets some biologists' needs without sacrificing key concepts that mathematicians and others deem important. (An example of this type is described in the sidebar on p. 20–21.)

A different challenge arises from the frequently voiced objection to splitting the life science stream (or any stream) in the first undergraduate year. Students entering college or university, so this argument goes, often have ill-formed and unrealistic ideas of their educational aims. Locking into one or another stream too early is said to reduce flexibility; supposedly this leads to greater problems later when students switch their major area of concentration.

Notwithstanding the seriousness of this concern, two important factors help reduce its impact. First, the kind of mathematics curriculum suggested above for biology students is valuable for students in a large number of other disciplines (with the possible exceptions of physics and some parts of engineering). Second, if we are to create a mathematically-talented workforce in biology, we

Science One
University of British Columbia

Science One is a 12-credit interdisciplinary first year program at the University of British Columbia (UBC) team-taught by four faculty members, one each from physics, mathematics, chemistry and biology. *Science One* was designed in 1992 by a committee of similar makeup. The goal of the program is to highlight connections between disciplines and challenge students to question, think, and be creative. First taught in 1993, the program rapidly became recognized as a greenhouse for talented students. The class consists of a group of approximately 70 students, self-identified via a general interest in science, but seeking a more challenging environment and an opportunity for intense interactions. (Their applications for admission to the program are ranked according to interests and accomplishments, social skills, hobbies, and experience—not only grade achievement.)

Faculty and instructors attend each other's lectures to gauge the level and opportunities for interconnections, to participate in discussions, and to enhance a feeling of community. The fact that several faculty and teaching assistants are dedicated exclusively to teaching *Science One* makes scheduling and coordinating material particularly flexible, with each discipline getting variable numbers of hours in a given week as required to coordinate or provide fundamentals before some interdisciplinary module or joint application is discussed. Team meetings once per week deal with technical issues such as who teaches what, how many hours each professor needs for the given material, and how to create modules that show connections.

For example, the mathematics faculty member may undertake to explain the wave equation using the force balance in a rope under tension; the physicist would then discuss the behavior of electromagnetic waves, and the biologist would follow up with lectures about photosynthesis. Before a unit on genetics, the mathematician would be asked to present one or two lectures about probability. The chemist's description of the structure of crystals would be accompanied by mathematical investigation of the packing of spheres or simple symmetries. In this way, numerous topics illustrate that science consists of many threads that together make up a richly woven cloth.

Students from *Science One* have the advantage of getting to know both peers and faculty very well; the group is often called a "community of learners." The group shares lectures, tutorials, discussion groups, and social events. Students hone communication skills through individual and team-based projects as well as in-class presentations. Expectations are very high, and not all students succeed. Those who do, make an impact on teaching across the Faculty of Science at UBC: Once taught how to question, think, and explore for themselves, they press for more analytic and concept-based teaching, more relevant applications, and deeper insights across all later years as undergraduates.

One alumnus, Aneil Agrawal, a Canadian Research Chair Assistant Professor in Zoology at the University of Toronto wrote that this "early year of exposure to the relevance of mathematics to biology has stayed with me and greatly influenced my choice of research career." As a graduate student in Biology, he authored prestigious papers in Nature and Science on the evolution of sex; his biological research has a strong mathematical component.

Science One also impacts teaching and learning in other respects. Faculty who rotate through this program export perspectives and lessons learned through this experience to the broader university community. As an example, a calculus stream for life science students (Mathematics 102/103) was developed by a team that includes Leah Edelstein-Keshet after her three-year stint teaching *Science One*.

Further Information:

J. A. Benbassat and C. L. Gass, "Reflections on integration, interaction, and community, The *Science One* program and beyond," *Conservation Ecology* 5(2) (2002) 26;
www.consecol.org/vol5/iss2/art26

www.science.ubc.ca/~science1/

Coordinated Science Program (CSP)
University of British Columbia

The Coordinated Science Program (CSP) at the University of British Columbia arose from a desire to share aspects of *Science One* with a broader group of first year students in a more economical setting. In CSP, class size consists of roughly 150-170 students, with an upper limit of 180. The syllabus of each of the core CSP classes—calculus, chemistry, biology, and physics—is not substantially different from that of the corresponding non-CSP first year course, and CSP faculty teach regular class hours.

Three features distinguish CSP from ordinary first-year offerings. First, students form a "cohort" that share a common section in each of their four basic courses and have some designated social/study space. Second, CSP faculty meet weekly to discuss strategy, find opportunities for linking concepts, or even adopt common guidelines about notation to help prevent confusion. Third, students participate in a two-hour weekly workshop, designed and implemented by instructors, where "modules" that stress interdisciplinary concepts and hands-on activities are carried out.

www.science.ubc.ca/~csp/

need to persuade students to take mathematics courses beyond the first-year level. Such planning may lessen the problems created by too-early specialization, since students who take more than one year of mathematics will, overall, gain a more well-rounded education.

In the second- and third-year courses, when the level of maturity of undergraduate students has increased, new topics can be introduced, including:

- Vectors and matrices—geometric properties as well as manipulations.

- Projections (dot products).

- Ordinary and partial differential equations (diffusion, waves).

- Nonlinear dynamics, steady states and stability.

- Eigenvalues and eigenvectors of linear systems, including geometric interpretation and applications of these to growth rates and age distributions in a population.

- Parameter variations and bifurcations.

- Mathematical modeling: how to read, understand, and be able to critically assess models in the biological literature; how to formulate and investigate models of increasing sophistication.

- The application of software and simple algorithms (such as cellular automata) to studying the behavior of models.

A significant number of these topics could be included in a course on mathematical modeling for biologists (or a general science-based modeling course for a wider audience) at the second- or third-year level. (An interesting technique in an interdisciplinary course is to pair mathematics and biology students in joint modeling proj-

Web Resources

Programs:

Science One (UBC):	www.science.ubc.ca/~science1/
Coordinated Science Program (UBC):	www.science.ubc.ca/~csp/
Integrated Science Program (UBC):	www.science.ubc.ca/~isp/
BioQuest (Beloit College):	www.bioquest.org/

Personal Web Sites:

Lou Gross (University of Tennessee):	ecology.tiem.utk.edu/~gross/
Eric Marland (Appalachian State):	www1.appstate.edu/~marland/
Joe Mahaffy (San Diego State):	www.rohan.sdsu.edu/~jmahaffy/courses.html
Bard Ermentrout (University of Pittsburgh):	www.pitt.edu/~phase/
Leah Keshet (Univ. of British Columbia):	www.math.ubc.ca/people/faculty/keshet/keshet.html

Software:

XPP (X-Windows Phase Plane):	www.math.pitt.edu/~bard/xpp/xpp.html
Berkeley Madonna:	www.berkeleymadonna.com/

ects.) More specific courses, targeted to biomedical students, should include:

- Mass action kinetics – its uses and limitations;
- Spatial and temporal analyses of systems: gradients, fields, flows;
- Transport phenomena – diffusion, convection, active transport;
- Feedback – positive and negative control;
- Waves and periodic phenomena – neural action potentials, cardiac dynamics, and excitable systems;
- Networks (genetic, biochemical, and signal transduction).

Some of these topics can also be introduced in biology courses in diagrammatic and intuitive contexts without elaborate mathematical buildup, and then reintroduced with the appropriate notation and precision in follow-up mathematics courses.

Faculty

While many of the above courses and programs are already under development at various sites, instituting change on a global scale is a daunting proposition that requires an accelerated pace. In this section we consider how faculty can be constructively engaged to meet this challenge. To do this, we examine several specific areas that need to be addressed, including time, funding, teaching credits, technical support, and related issues.

Faculty members are rewarded within the university and college system for producing output valued by their employer. In many research universities, a strong research publication record is essential for promotion and tenure, while good teaching is important but not sufficient. In this environment, it makes sense for faculty to focus most efforts on research and optimize the efficiency of teaching. Teaching familiar material in traditional ways makes for such efficiency. Changing these ways and revising curricula takes time and energy. Ultimately, given time, market forces will lead to selection of young new interdisciplinary faculty who are well-equipped to create the kinds of new training programs espoused in this report. However, for more rapid change, we need to enlist the active support of current faculty whose training was discipline-specific. To encourage such change, we need new incentives that address the needs of the professoriate and motivate innovation in teaching.

Time. Most university and college professors are overcommitted in teaching, research, and service to their institutions. They need time to develop new materials and approaches to teaching, to participate in committees dedicated to innovations in teaching, and to learn about material that others have developed (e.g., at conferences and workshops). The required time can be released from current obligations or added as paid summer work. Recognition of time and effort (for example, through merit increases or faculty achievement awards) could also be part of an incentive package. Adjusting teaching assignments to support development of new courses or participation in team teaching is also important.

Departments, faculties, and universities need to confront this issue directly. Scientific societies can help by recognizing colleges and universities for innovative teaching ventures. For example, an annual feature in Science about innovations in undergraduate bioscience education could fuel the impetus to bring about these changes. Universities welcome external recognition and publicity in their competition for potential students.

Funding. Innovation is an expensive undertaking, and many efforts require funding at several levels. We mention a few of these below:

Technical aspects of course preparation (developing websites, preparing illustrations, formulating and typing solutions to exercises) can be time-consuming and unappealing. With help for such mundane tasks, many courses could be transformed from concept to reality. Funding for these technical aspects of course preparation is an important factor at the departmental level; it is an essential factor if the efforts of a few innovators are to be disseminated to others. (At UBC, a Faculty of Science "Skylight Fund" awards small stipends of $1000–$4000 to faculty on a competitive basis to develop and disseminate course material.) Teaching assistants and instructors can also be recruited to provide the technical help needed in course preparation, albeit at somewhat higher cost.

Financial support for retraining at workshops and for time spent at such workshops can greatly increase the number of individuals recruited to disseminate innovative approaches. Some grants have enabled the scientific and mathematical societies to award expenses and partial travel costs to faculty attending workshops, on a competitive basis, given clear intent to put the experience gained to use in teaching. This model is worth emulating.

Summer salaries or prizes for excellence of educational innovation can further engage leaders at the interface of mathematics and biology (who would otherwise be focusing on innovation in research or business). This funding need can be addressed by NSF, NIH, Howard Hughes, and other funding agencies.

Unfortunately, most universities in the United States—both public and private—are experiencing crushing funding reductions that have caused some to retreat from innovations in teaching and return to teaching traditional material in large classes. Since these conditions will likely persist for several years, funding at the national level—from government agencies and private foundations—is crucial to any significant near-term introduction of additional mathematics into the biology curriculum.

Teaching Credits. Many departments are funded by the university in proportion to the number of majors they train. Hence team-teaching or teaching for other departments is not easily justified. To alleviate this restriction, some flexibility is needed at the level of universities and faculties. Credits could be awarded for teaching across disciplines, as well as for new course development. A reduction in teaching loads can lead to more time rethinking the curriculum and developing new courses.

Material. At present, efforts to create effective biological and mathematical programs and interdisciplinary courses are in great abundance, albeit at formative stages. These communities are poised to undertake the challenge, but to enable this process, they need the support of teaching-related material such as:

- Short, well-designed modules that can be used in biological or mathematical courses to increase their interdisciplinary content. Some such material exists, although it is not well known or easily accessible. Some examples can be found at Lou Gross's website at the University of Tennessee (see web resources on p. 69).
- Clearinghouses for these modules with simple effective indexing and descriptive summaries. Ease of access would be an inducement to adopt, adapt, and improve this material.
- Reviews of what works and what fails; discussions of experiences and experiments in teaching.
- Simple modeling software that runs on personal computers. Examples currently available include XPP (from Bert Ermentrout at the University of Pittsburgh) for advanced students, the Berkeley Madonna, and various BioQuest modules for biologists. (See the sidebar on p. 25 for web site addresses.)

Above all, a variety of texts are needed for the emerging life-science mathematics and quantitative biology. This need is most acute at the freshman level. A text has the advantage of offering a self-contained summary that

can be used by professors who are not yet fully familiar with the entire landscape. In the past, well-written books have served as catalysts for new courses (Birkhoff & MacLane, 1977; Kemeny, Snell & Thompson, 1956; Strogatz, 1994). At present, the number of texts suitable for first year offerings of mathematics for biologists are limited (but see Adler, 1997 and Neuhauser, 2000 for two examples.)

Space and Equipment. Antiquated classrooms make use of computer-aided technology almost impossible. The time required to connect data-projectors, laptops, and intra- or internet ports can be very discouraging; this factor alone can prevent many faculty, who are always short of time, from making the effort to change. It can also facilitate (or prevent) the use of a vast set of online resources that could stimulate and challenge students during the course of a lecture. Thus updating classrooms and their equipment are crucial steps in meeting the challenge. Further, computer laboratories equipped with personal computers and a variety of modeling software are vital in both undergraduate mathematics and biology. In some colleges, classrooms are equipped with individual personal computers, and instructors engage the class with interactive computation during the lecture hour (see the sidebar on p. 23). Even a more modest scenario, with demonstrated computation that students follow up at home or in the lab can accomplish similar goals.

Research Support. Many faculty with innovative research programs can be enticed to extend their reach to innovative teaching, given assured support of their continued research. For example, funding a talented postdoctoral fellow can help to maintain an active research program while assisting in teaching and course design.

Intangibles. Here we list some factors that are not quantifiable in direct financial or material gain, but that play decisive roles in facilitating changes in the mathematical training of biologists.

Departments should strive for a greater flexibility about what a student "has to have" to complete requirements for the bachelor's degree. Broad interdisciplinary major programs are more likely to produce the scholars needed for the future. At the same time, we must balance breadth with depth and avoid diluting the content of courses to an extent that leaves large gaps. Reconciling these competing goals will take some creative thought.

At the university level, scholarly activity will need to be more broadly defined if rewards are to follow effort in educational innovation. Currently, promotion and tenure decisions are based on the triad formed by scholarly activity (usually interpreted as research), teaching, and service. Education innovation could be rewarded as one type of scholarly activity (Boyer, 1991). A further need is recognition of the importance of interdisciplinary work. Finally, flexibility in the funding of departments to encourage outreach and cross-disciplinary ventures will be important.

Professional societies can contribute to educational reform by highlighting successful programs, increasing the visibility of success stories, and bringing the issue of education to the attention of their members. Prizes for innovation and for shared and disseminated material can motivate individuals and their institutions to participate, and to cooperate rather than compete in these aims.

Finally, part of the responsibility for reform lies with the faculty itself, regardless of incentives and external resources. Faculty members have to learn to communicate with their peers in other disciplines, to overcome barriers of culture and language. This takes time, patience, and recognition that paradigms are specific to each discipline.

Students

To make innovative education work for students, attention must be paid to student needs. Reformed curricula and faculty commitment can meet only part of the challenge. Other necessary ingredients pertain to students—their intentions, desires, and commitment. We discuss some of these below.

Helpful Information. Students are often unaware of how decisions about courses and programs at early stages of their education affect their later career options. This lack of awareness starts well before college years: many lack information about the changing careers in the life sciences and about the universal importance of mathematics, biology, and computational skills. A much better system of advising and clearer guidelines about course planning should be made available to students in grades 6-12, and to undergraduates at all stages. Informative pamphlets addressed at high-school students and teachers can highlight the career paths of leading scientists and explain what they had to learn to achieve their goals. A better system of undergraduate advising is also essential.

Some students chose life science majors out of fear of the quantitative sciences. Others do so with the intention of applying to medical school. Many of these students are unaware of the universal need for increased quantitative and mathematical skills (and the relatively low probability of acceptance to medical school). Such misconceptions lead to loss of vital opportunities to acquire these essential skills, and consequent loss of time in educational advancement. Biology faculty, including professors who were trained in traditional biology, should strongly encourage students to take (and stick with) appropriate courses in mathematics, physics, and computer science.

Inclusive Environment. Recognizing the enormous diversity of experiences, cultures, and aspirations of students, the educational environment in the early years of college or university education must be both inviting and inclusive. At the same time, it should maintain high expectations for everyone. One great strategy is to "infect" talented young people with the excitement of science. We need to communicate the appeal of a career in science. The vast majority of undergraduates have had little or no experience with real science prior to their university education, and consequently, have no idea of the fulfillment and sense of adventure that it can provide.

Exposure to positive, early research experiences during the first year or two of undergraduate education (or even earlier) can be a terrific incentive. Further motivation can come from positive role models and mentors (both student peers and faculty) who provide examples of what can be accomplished. In many institutions, second-year alumni of mainstream first-year courses serve as lab or section leaders. Their presence provides positive reinforcement for incoming students in these courses.

Learning communities—groups of students who share several courses, residential life, or study groups—enhance the retention of students. Facilitating such communities is well worth the effort, especially on large campuses. They also offer opportunities for assigning challenging team projects and encouraging work in groups. Not only do these experiences offer opportunities for social interactions and favor retention, they also introduce students to teamwork prevalent outside academe.

A final consideration is the involvement of students in research experiences, modeling contests, evaluation, and even design of courses. Such ventures promote entitlement, ownership and commitment, and empower students to take charge of their educational experience. By providing such environments, today's faculty can contribute to the creation of the biological, mathematical, and interdisciplinary workforce of the future.

Finally, in order to sustain student interest, courses designed for the "new biology"—whether in biology, in mathematics, or in computer science—need to stress conceptual understanding, not just techniques and facts.

They need to introduce applications, hone problem-solving skills, and motivate thinking and intuition—rather than mere rote memorization. These new courses need to encourage students to think about and formulate problems in a variety of ways, and not simply learn to recognize patterns in solving standard exercises. In short, these courses need to expose students to the excitement of scientific research early in their careers. All these student-oriented goals provide grand challenges for designing new curricula.

Bibliography

Adler, F.R., *Modelling the Dynamics of Life: Calculus and Probability for Life Scientists*, Brooks/Cole, 1997.

Birkhoff, G., and S. MacLane, *A Survey of Modern Algebra*, Prentice Hall, 1977.

Boyer, E.L., *Scholarship Reconsidered: Priorities of the Professoriate*, Carnegie Foundation for the Advancement of Teaching, 1991.

Kemeny J. G, J. L. Snell, and G. L. Thompson, *Introduction to Finite Mathematics*, Prentice Hall, 1956. (Available at: math.dartmouth.edu/~doyle/docs/finite/cover/cover.html)

Neuhauser C., *Calculus for Biology and Medicine*, Prentice Hall, 2000.

Strogatz S.H., *Nonlinear Dynamics and Chaos*, Addison Wesley, 1994.

Computer Science and Bioinformatics

Paul Tymann et al.

This paper was a collaborative effort co-authored equally by the following individuals:

Clare Bates Congdon, Colby College, ccongdon@colby.edu;

John Dougherty, Haverford College, jdougher@haverford.edu;

David Evans, University of Virginia, evans@cs.virginia.edu;

Mark LeBlanc, Wheaton College, mleblanc@wheatoncollege.edu;

Joyce Currie Little, Towson University, jlittle@towson.edu;

Jane Chu Prey, National Science Foundation, jprey@nsf.gov;

Vojislav Stojkovic, Morgan State University, stojkovi@morgan.edu;

Paul Tymann, Rochester Institute of Technology, ptt@cs.rit.edu.

Bioinformatics is an interdisciplinary field that requires teams of experts in biology, chemistry, computer science, and mathematics working together to analyze the terabytes of biological data that are rapidly accumulating in laboratories around the world. Unfortunately, since different scientific disciplines do not share the same scientific language, forming an effective bioinformatics team is not as simple as just arranging for a group of scientists to work together on a common problem. An effective bioinformatics team requires two kinds of experts: (a) computer scientists and mathematicians who are excited about and understand the basic biological processes being studied, and (b) biologists and chemists who are interested in and familiar with the basic computational processes used to manipulate and analyze biological data. Working as a team they can leverage different areas of expertise to form a powerful intellectual partnership that uses key insights from multiple disciplines to solve problems that cannot be solved by any single discipline.

Teamwork in bioinformatics is difficult for many reasons, not least because few computer scientists have studied much biology, and most biologists' exposure to computer science has been through introductory courses that focus on programming rather than the science of computing. Our goal in this paper is to help all those involved in "meeting the challenge" understand what computer science is and how it contributes to the study of bioinformatics. An appreciation of computer science as an intellectual discipline will help to set the stage for successful interdisciplinary collaborations.

This paper begins with a description of the discipline of computer science, focusing on some key insights that a computer scientist brings to the study of bioinformatics. It then describes a minimum set of computing skills and knowledge that a non-computer scientist should have in order to work effectively with a computer scientist. The paper concludes with a discussion of a few of the academic programs being developed to educate the next generation of bioinformaticists who will have a basic understanding of all the disciplines involved in bioinformatics.[1]

Computer Science and Biology

Computer science is the study of information processes. It is not a physical or natural science since it deals pri-

[1] The authors would like to thank Betsey Dyer, Jim Fink, Rhys Price-Jones, Carl Leinbach, Michael Schneider, and Gary Skuse who read early drafts of this paper and provided valuable comments and criticisms.

marily with abstract ideas such as numbers, graphs, strings and procedures. In this regard, it is more like mathematics. The key difference between computer science and mathematics is that pure mathematics is declarative (it deals with what is knowledge) whereas computer science is imperative (it deals with how to knowledge). For example, Euclid was not a computer scientist even though he developed a systematic method for computing the greatest common divisor. He simply revealed his method by describing it. Euclid did not consider how his method could be implemented on a computing device, nor did he reason about the properties that an implementation of his method might exhibit. Augusta Ada Byron (1815–1852), on the other hand, was a computer scientist. She thought about implications of methods for computing (e.g., those associated with Charles Babbage's analytical engine) and found those methods, their implementations, and the properties of their implementations intrinsically interesting.

Computer science focuses on understanding how to describe procedures (that is, precise methods for manipulating information) and how to reason about the information processes they encode. A sibling field, computer engineering, is about economically producing computing machines that can execute procedures efficiently and reliably. The combination of computer engineering and computer science has produced many practical tools that are of great use in biological investigations. Perhaps the most visible example is the essential role played by computers and computer algorithms in the human genome project.

Biology is the study of life. Life emerges from complex interactions among physical and chemical processes that are described and encoded using information processes. During the twentieth century, biologists came to understand many aspects of those physical and chemical processes and learned how information is encoded in chemical structures (DNA). We have not yet learned how the complex physical, chemical and informational processes transform a single cell containing DNA into a complex, robust, living organism. The human genome project succeeded in producing a nearly complete human genome sequence. We know enough about chemistry to collect and organize all the information in a genome, but do not yet know enough about the processes that DNA encodes to use all that information in a useful way.

Key Insights from Computer Science

Computer science is a young field that only began in earnest in the 1940s (Byron's work in the 1840s was vir-

tually ignored for a century.) The first university degree program in computer science did not appear until the 1960s. However, since that time computer science has grown rapidly as a field and has had remarkable impact on our society, affecting all aspects of modern society from automobiles to zoos.

Some of the most important insights in computer science concern ways of describing and predicting the behavior of information processes. Among them are:

- that things can be defined in terms of themselves (recursive definitions);
- that procedures and data are not different kinds of things, but the same;
- that abstraction simplifies the understanding, and design, of complex computing systems.

Although these ideas appear simple, they have profound implications that can take many years to fully appreciate. They also have direct relevance to biology. For example, biology is recursive at many levels. Most cells in an organism contain the DNA necessary to produce a complete, similar organism; at a higher level the branches of a tree, the blood vessels in a circulatory system, and the social organization of an insect colony are all deeply recursive, as is the process of evolution by natural selection. The interchangeability of procedures and data is also apparent in a genome, which is both a chemical structure and a program for producing an organism. Finally, biology groups related things into units in hierarchical ways that can be viewed as abstractions. We have genes and chromosomes, not just unstructured sequences of nucleotides. Organisms are made from organs, which are made from tissues, not just amorphous masses of cells.

Computability and complexity are two computer science concepts of particular interest to biologists. These concepts provide a way to reason about the behavior of information processes, making it possible to determine whether one can compute something with an algorithm, and how much time and space will be required to solve a problem for a given size input:

Computability. Certain problems are decidable; others are not. A decidable problem is one that can be solved by an algorithm, that is, a procedure that is guaranteed to always terminate in a finite amount of time. Decidability is independent of the machine on which the algorithm is to be implemented. All conceivable computing machines are deeply equivalent: except for arbitrary limits on speed and memory, any non-trivial computing machine is as capable as any other. A Turing machine, an abstract representation of a computing device developed by Alan Turing [Turing01], is able to simulate every other computing machine.

Complexity. The amount of time and space (memory) required to evaluate a procedure for a particular input can be predicted based on the size of the input. Some problems can be solved in a reasonable amount of time even for large size input; the class of all such problems is called P (for polynomial time). Another class of problems is called NP (non-deterministic polynomial time) if there is an algorithm that can check whether a proposed solution is correct in a reasonable amount of time, but for which we have no known algorithm that can find a solution in a reasonable amount of time. An important subclass, NP-complete, consists of the hardest problems in NP.

All NP-complete problems are equivalent in a strict algorithmic sense. That is, if a fast (i.e., polynomial-time) algorithm is found for any one of them, then a fast algorithm can readily be found for all of them. On the other hand, if it can be shown that no polynomial time algorithm exists for any single NP-complete problem, then no fast algorithms exist for any of them. To date, the only algorithms known for NP-complete problems would take millions of years to execute for modest size inputs even on the fastest computers.

Using these insights a computer scientist is able to reason about the information processes associated with biological procedures. Consider, for example, the problem of sequence reassembly. A virologist may wish to determine the sequence of a 20,000 nucleotide (nt) viral genome. Sequencing machines are accurate only for nucleotide molecules of less than about 700 nt, so to map the entire viral genome the virologist may produce a collection of overlapping fragments, each of lengths 300–500 nt, that together cover the entire genome. The challenge of sequence reassembly is to determine the original 20,000 nt genome by properly arranging the overlapping fragments whose nucleotide sequences (strings) have already been determined by the sequencing machine.

It is easy to find a superstring that contains each fragment as a substring: simply concatenate all of the fragments. But to determine the proper order of the fragments, they should not merely be concatenated but overlapped where common sequences occur. Maximum overlapping will yield a minimal superstring. This shortest common superstring of all the fragments is a likely candidate for the genome the virologist seeks.

Unfortunately, the task of finding the shortest superstring is NP-complete. The only known algorithms that solve it definitively require time that increases exponentially with the number of input strings; no known algorithm will produce the shortest common superstring in a reasonable amount of time for any realistically sized genome. There are, however, algorithms that will find superstrings no longer than, say, 5% greater than the shortest common superstring and that might be good approximations to the original genome. However, in most cases millions of different superstrings will be close to the minimal length, differing from one another in the order of the fragments used to identify the nucleotide sequence. Because most computer scientists believe it is impossible to find a fast solution to an NP-complete problem, the best we can expect—as illustrated in sequence reassembly—is to develop fast algorithms that find approximate (but not necessarily optimal) solutions.

When a problem is NP complete, and therefore prevents development of an algorithm that will definitively calculate the optimal solution, computer scientists turn to heuristics, which are rules of thumb about how to find the best solution. The term "heuristic" may be applied to the process of evaluating proposed solutions to a problem (heuristic evaluation functions). For example, when doing sequence alignments, a scoring matrix is a rule of thumb that tells us which alignments we consider more likely; different matrices are preferred according to different technical bases. The term "heuristic" is also often applied to the process of searching for candidate solutions (heuristic search methods). For example, BLAST employs a heuristic search for sequences in a database (see the sidebar on p. 34). One of the rules of thumb invoked in this search is to first look for short sequences in common with the query sequence. This heuristic enables it to return sequences from the database in a reasonable period of time.

Many biologically relevant problems are NP-complete. When the input size is small, it is possible to solve these problems by trying every possible solution and identifying the best one. However, when the input size is large, heuristic methods need to be invoked. As another example, phylogenetics involves a search among all possible evolutionary trees to find the tree with the smallest number of character transitions required to explain the observed variations in the species or sequences. It is computationally infeasible to search through all possible trees for more than 20 species or sequences, so computer scientists will turn to heuristics to guide the search process.

Understanding how to reason about decidability of problems and complexity of procedures can change fundamentally the way biologists think about problems.

BLAST

The BLAST programs (**B**asic **L**ocal **A**lignment **S**earch **T**ools) are algorithms for searching nucleotide and protein databases for optimal local alignments. The BLAST programs break database sequences into fragments ("words"), and initially seek matches between fragments. Word hits are then extended in either direction in an attempt to generate an alignment with accuracy exceeding a specified threshold. Since the BLAST algorithm detects local as well as global alignments, regions of similarity embedded in otherwise unrelated proteins can be detected. Both types of similarity may provide important clues to the function of proteins.

Further Information: www.ncbi.nlm.nih.gov/Education/BLASTinfo/information3.html

These insights make it possible to determine which problems can be solved by using more computing power and which require either fundamental breakthroughs or settling for approximate solutions.

Opportunities and Challenges

Traditionally, advances in biology (as in most scientific studies) occur most frequently through deep investigation of a narrowly defined problem. Many years of intense education and concentrated research are necessary before progress is realized. As the previous discussion illustrates, some recent advances in biology require application of concepts that have traditionally been part of computer science, not biology [CUB03]. This changing reality reveals one aspect of the "challenges" of bioinformatics: many biologists may not be prepared to incorporate necessary concepts from computer science into their research or teaching.

There are at least two approaches to this challenge. The first is to extend research groups to include experts in computing. Collaboration among a group of individuals with different expertise could limit the extent and depth of understanding of computer science needed by biologists in order to speak to their computer scientist partners. Computer scientists often develop software in an interdisciplinary environment, thus have identified special issues (e.g., vocabulary problems [Fur87]) and potential responses that arise in such contexts.

Most biologists, however, have not had equivalent cross-disciplinary experiences. Being good scientists, many biologists—perhaps most—believe that one cannot make sense of data without a complete understanding of the underlying phenomena. "Until you've held a pipette in your hand, or run a gel or PCR [polymerase chain reaction], you can't understand what the failure modes are going to be," explains Mark Perlin, chief executive officer of the genomics company Cybergenetics. "The basic mentality clash between data-driven scientists and model-driven computer and math people stems from world views so different that all too often they can't have a conversation that makes sense." [Wick00]

The second approach involves extending the capabilities of biologists to include necessary skills from computer science [ACM01]. This approach can only be implemented as a long-term strategy, since it requires biology researchers to actively study computer science while simultaneously working on research projects—a demand that few could be expected to meet in a short period of time. Students—both graduate and undergraduate—are more likely than current researchers to readily develop both wet-lab and computer science expertise [Wick00]. For example, the University of Washington has designed a new curriculum that explicitly views biology as an information science [UW03]. Variations can be found in other places (e.g., [Sus95, Whea03, Ohio03]). It will take time for graduates of these programs to develop and contribute as "new biologists."

In both of these approaches, the scientific communities in biology and computer science need to cooperate in addressing the issues of collaborative research. In order to maximize the opportunities available, multi-disciplinary preparation must be encouraged at various stages from high school through graduate school and on into funded research projects. To increase the chances of success, strategies to overcome obstacles need to be diverse, ongoing, and adaptable. Real problems need to be solved by people with the vision to see these opportunities and the willingness to embrace new ideas.

Computing Skills for Non-Computer Scientists

One of the challenges in describing the computing skills required to be an effective member of a bioinformatics research team is to distinguish between fundamental understanding of computing and the ability to use computers that a society expects of educated individuals. At least four major movements have influenced the definition of computer skills expected of all: computer literacy, information literacy, information fluency, and "computing across the curriculum." These movements have created a gradual but persistent force to prepare all members of society for the information age.

The *computer literacy* movement was initiated in the 1960s even before personal computers were available. Over the years, change has occurred based on the recommendations of various computing societies, government agencies, and many educator groups. More recently, computer literacy, as defined by information technology professionals, is knowing how to use personal productivity products such as word processing, spreadsheets, presentation graphics, and database software on a personal computer. From this perspective, computer literacy is the ability to successfully utilize the computer as a tool to enhance productivity. Others argued that it should also include an understanding of the societal impact of computers.

The *information literacy* movement of the 1980s and '90s became the major initiative to define what college graduates should know. Like the computer literacy movement, its emphasis was on use of computers, not on the concepts of computer science. Information literacy combined the tool-oriented skills of the computer literacy movement with a focus on how to find and use information for research. This focused primarily on the use of library facilities for research and included topics such as learning how to do research in the library, using a web browser, awareness of established library databases, and using digital libraries.

A 1999 report by the National Research Council (NRC), "Being Fluent with Information Technology," caused many university and college systems, as well as some secondary school systems, to take a fresh look at what all students should know about computers and computing [CITL99]. This movement can be considered an update to computer literacy, and is called information fluency. Included in information fluency is the capability to use computer software for personal and professional skills in word processing, electronic spreadsheets, presentation graphics, and databases. Further, most information-fluency goals include the capability to use the Internet, to be able to use web browsers, and to conduct web searches. In addition this report recommended that individuals have some skill in knowing what computer programming is and how to set up a problem for programming. This NRC report urged that everyone develop the capability to apply these skills and concepts to everyday life.

Finally, numerous curriculum reports of the Association for Computing Machinery (ACM) define the various specialties within computing and information sciences for secondary schools, two- and four-year colleges, and graduate programs [ACM01]. Many of these reports include a category known as computing for other disciplines, or computing across the curriculum. These reports, and others by other associations, recognize the need for specific computing skills for a variety of disciplines. Without exception, they recommend multi-disciplinary teams to further define the needs for each discipline.

While these various skills are vital for anyone to be successful in today's society, these movements do not focus on computing concepts but on basic computer literacy and the use of computer technology. Since this is the type of knowledge that most individuals know, many mistakenly equate computer literacy with computer science. Although every scientist should be well versed in basic computer literacy skills, many are unfamiliar with the basic concepts underlying computer science.

Key Computing Concepts for Biologists

This section identifies areas of study for biologists seeking to increase their knowledge of computing, including program design, algorithmic analysis, database systems, and advanced computing techniques. Biologists' progressive study of and exposure to these areas will greatly facilitate effective and creative collaboration in bioinformatics. (Some of these concepts are based on those recommended by the Computer Science Panel of the NRC Study Group [CUB03].)

Algorithmic Analysis. The ability to develop algorithms to solve problems, the appropriate use of heuristics, and the ability to analyze and predict the performance of algorithms are among the basic concepts of computer science essential to bioinformatics. Biologists do not need to be proficient in the development of algorithms, but they should understand the algorithmic process. This parallels the computer scientists' need to understand how PCR works. Computer scientists need not master the process of actually running a gel, but they should be aware of the process so that they understand its limitations as well as the kinds of problems that may arise. Biologists should be familiar with the methods used to develop and express and analyze algorithms.

Program Design. It is important that biologists working in bioinformatics understand the basics of software development. Biologists should be able to develop a simple algorithm and know how to implement it in a programming language. This knowledge might be obtained from the central programming courses (CS1 and CS2) offered at the beginning of the undergraduate computer science curriculum, but these courses rarely dwell on problems from biology. A separate course that focuses on bioinformatics may be more appropriate.

Database Systems. The explosive growth of biological data has created demand for the use and development of techniques to store and retrieve these data. While simple data structures are sufficient to store and retrieve small data sets, more sophisticated approaches are required to process large amounts of data. Large data sets require especially efficient storage and retrieval mechanisms so that queries are expedient and versatile. In order to achieve this expediency and flexibility, the internal representation used to store and retrieve the data must be carefully designed. Since a vital part of almost all bioinformatics research involves working with databases, "new biologists" will need a basic understanding of the tools and techniques used to manage information on a large scale.

Advanced Computing. Computer scientists use many techniques to solve difficult problems. Of these techniques, machine learning, data mining, and high-performance computing play an important role in bioinformatics research.

- Machine learning is the study of programs that improve their performance with experience, for example, those that induce patterns in a dataset and use these patterns to make predictions about future data. Because the learning process is inductive, machine learning is an inherently heuristic process, seldom guaranteed to have found the definitively best patterns that describe a dataset or to predict unseen patterns 100% correctly. Machine-learning techniques are commonly used in bioinformatics tasks such as the prediction of protein folding, the location of potential genes within a sequence, or the identification of microarray expression patterns that are associated with a particular disease.

- Thousands of online databases contain a wide variety of biological data recording the results of millions of experiments that have been performed in laboratories around the globe. Several new discoveries in bioinformatics have been made just by analyzing the contents of different databases and identifying relationships in the information that they contain. Data mining is a technology that uses pattern recognition and machine learning to analyze databases in order to identify relationships in the data stored in these repositories and to discover new information based on these relationships.

- As more and more biological data is collected and archived, the computational power required to process this data has increased significantly. As a result a typical single-processor computer does not have the power to process this data in a reasonable amount of time. Many bioinformaticists are now turning to parallel- and distributed-computing solutions to solve computationally challenging problems. As high-performance computing becomes more mainstream within the bioinformatics community, biologists will need to understand parallel and distributed systems, determine the type of high-performance hardware and software that would be required in their work, effectively evaluate commercially available hardware and software systems, and have the knowledge required to work with computer scientists to develop software that takes advantage of high-performance systems.

It is not essential that biologists become experts in any one of these areas, but rather that they become familiar with the basic concepts in each. The goal is not to create a clone of a computer scientist but to arm biologists with fundamental knowledge that will enable them to understand and discuss issues, problems, and limitations as part of an interdisciplinary process. This is similar to the need for computer scientists to understand the process involved in collecting data from a micro-array experiment. It is highly unlikely that a computer scientist would ever have to be proficient in this wet-bench technique, but it is vital that they understand the issues, problems, and limitations of the technique. Moreover, they need to understand how certain situations can produce incorrect results so they may take this into account when developing algorithms to analyze the data. So too do biologists need to understand fundamental issues in computer science.

Examples of Undergraduate Bioinformatics Programs

Not surprisingly, the introduction of bioinformatics has proceeded differently at different institutions as resources, goals, and staffing vary considerably. At some institutions encouragement and resources for bioinformatics comes from the administration; at others, the push comes from individual professors. Approaches to introducing bioinformatics to undergraduates range from the infusion of bioinformatics concepts into existing courses to the creation of entirely new degree programs [Dyer02].

Full degree programs for undergraduates are emerging at many large universities (e.g., Rochester Institute of Technology, Rensselaer Polytechnic Institute, Wright State, and the University of California Santa Cruz) and at a small number of liberal arts colleges (e.g., Canisius, Ramapo). A growing number of liberal arts colleges are developing a minor in bioinformatics or a tailored concentration within a major that provides program-specific recommendations depending on the major. A significant number of colleges and universities are team-teaching across interdisciplinary boundaries or "linking" two related courses; examples include Colby, Quinnipiac, Union, Wheaton, and Williams. Some schools have relevant expertise within the academic staff; others introduce bioinformatics via retraining existing faculty or hiring new faculty. What follows is a description of four different examples of bioinformatics offerings and the paths used to create them.

Colby College is a liberal arts college in Maine with 1,800 students. Efforts to introduce bioinformatics to the curriculum are supported by the administration with the help of grants from the Howard Hughes Medical Institute (HHMI) and the Biomedical Research Infrastructure Network (BRIN). The introduction of bioinformatics at Colby is proceeding along several fronts.

First, Colby offers an interdisciplinary course, co-taught by professors in biology and computer science; students survey the breadth of bioinformatics during the first half of the semester and work on original research projects during the second half. The course is run as a seminar with few lectures. Instead, students are responsible for leading discussion and answering each other's questions. Weekly computer labs during the first half of the semester give students hands-on experience using systems and languages commonly used in bioinformatics. Both professors are present at all class meetings to guide discussions and to help answer questions.

Second, both the biology department and the chemistry department have added bioinformatics modules to existing courses. Third, both the biology department and the computer science department have designed concentrations in bioinformatics to be added to the respective majors.

Haverford College is a liberal arts college near Philadelphia with 1,100 students. Support for the introduction of bioinformatics comes both internally from the college and externally from an HHMI grant. Beginning

in 2000 the HHMI grant funded a four-year seminar series for Haverford faculty intended (a) to address faculty expertise in a broad range of areas of scientific computing and (b) to expand and deepen the ways in which faculty integrate an understanding of the relationships between science and society into their teaching [Hav00]. The first year theme was "Computing Across the Sciences;" one of the modules in this seminar involved genome sequencing.

The second year was dedicated entirely to bioinformatics, with three professors from the University of Pennsylvania conducting weekly seminars and workshops [Ewen01]. The goal of the second year was to expose faculty to topics in bioinformatics that could be injected into existing courses, or inspire the development of new ones [Hav01]. Bioinformatics now appears in several places in the curriculum, including "Computing Across the Sciences," the introductory computer science courses (CS1 and CS2), "Mathematics for Biology," "Seminar in Genomics," and "Scientific Computing."

Wheaton College is a liberal arts college of 1,500 students located in eastern Massachusetts. For four years faculty in biology and computer science have team-taught and linked courses in computer science and biology. "Linked" courses are independent courses (e.g., Algorithms and Genetics) that share guest lectures, collaborative labs, and student-paired capstone research projects. The labs enable students to collaborate on software specifications and designs. By strategically selecting the courses to link, this model reaches 100% of the majors in computer science (Algorithms is a required course) and over 70% of the majors in the life sciences. The National Science Foundation and the Wheaton administration have funded course development and research that involve students and faculty from biology, mathematics, and computer science. Faculty in these departments plus chemistry and statistics are currently developing a major in bioinformatics.

Rochester Institute of Technology (RIT) is a mid-sized university in upstate New York with an undergraduate population of approximately 12,000 students. Support for a new bioinformatics degree program came both from within the institution and external grants. The curriculum was developed through the efforts of an interdisciplinary group of RIT faculty (representing biology, biochemistry, statistics, and computing sciences) and was validated by oversight from an external advisory board consisting of scientists from industry and academia. It is a fully integrated curriculum that requires students to take courses in biology, computer science, mathematics, and information technology that are taught in two different Colleges. In addition to traditional course work, all students are required to complete at least one 10-week cooperative-education experience designed to provide them with real-world experiences in bioinformatics in either industrial or academic research laboratories.

The majority of the courses required for the new program were already being offered at the Institute. For example, all the required computer science courses are already part of the computer science major (CS1, CS2, CS3, and Programming Language Concepts). However, twelve new courses were designed and are in the process of being implemented. Two of these, Introduction to Bioinformatics Computing I and II, focus on algorithms that are currently being used in the field of bioinformatics; a third, High-Performance Computing for Bioinformatics, will focus on the use of parallel and distributed computing techniques to solve computationally challenging problems. Teams consisting of both biologists and computer scientists teach many of the courses in the program.

Graduates of this program will have a strong foundation in biotechnology, computer programming, computational mathematics and statistics, and database management. Accordingly, they will be well prepared for careers in the biotechnology and biopharmaceutical industries or for graduate programs leading to advanced degrees.

Summary

Developing an interdisciplinary team of scientists that can work together effectively to achieve some of the promises of bioinformatics is a challenge. In order to be effective computer scientists must understand the basic biological concepts being investigated, and biologists must understand the basic computational process being used to manipulate the data. In order to be successful, each member of the team must have a basic understanding of each of the disciplines represented by other members of the team.

A computer scientist brings to the team the ability to implement computational models and can provide the formal reasoning required to understand the computational processes that are used to analyze the biologists' data. A biologist brings an understanding of the biological processes at work within living organisms, coupled with the ability to collect data that reflect the way these processes work.

The key questions for science in the twenty-first century will involve understanding the deep and complex interactions among biological, chemical, informational and physical processes. The current generation of scientists was not educated in ways that provide sufficient understanding of (or even expectations for) these kinds of connections. They (we) were educated within the boundaries of historically defined disciplines and have had to work against tradition to break these bonds. The next generation of scientists, educated in biology, chemistry, computer science, and mathematics, will be at home with deep interaction between these fields. Their research will transform our understanding of living organisms and is a worthwhile challenge to meet.

References

[ACM01] Joint *ACM/IEEE-CS Task Force on Computing Curricula, Computer Curricula 2001 for Computer Science*, (Dec. 15, 2001). www.computer.org/education/cc2001/ final/index.htm.

[CITL99] Committee on Information Technology Literacy, National Research Council, *Being Fluent with Information Technology*, National Academies Press, Washington, DC, 1999.

[CUB03] Committee on Undergraduate Biology Education to Prepare Research Scientists for the 21st Century, National Research Council, *Bio2010: Transforming Undergraduate Education for Future Research Biologists*, National Academies Press, Washington, DC, 2003.

[Dyer02] Dyer, B. and M. LeBlanc, "Incorporating Genomics Research into Undergraduate Curricula." Cell Biology Education, 1(4), Winter 2002, 101–104.

[Ewen01] Ewens, W.J., and G.R. Grant, *Statistical Methods in Bioinformatics: An Introduction*, Springer, New York, 2001.

[Fur87] Furnas, G. W., T. K. Landauer, L. M. Gomez, and S. T. Dumais, "The vocabulary problem in human-system communication," *Communications of the ACM*, 30:11 (November 1987) 964–971.

[Hav00] Haverford College, "Computing Across the Sciences." www.haverford.edu/ biology/cats/hhmi.htm.

[Hav01] ——, Bioinformatics: A Faculty Development Course. www.haverford.edu/biology/ facdevelop.htm.

[Ohio03] "Quantitative Biology Institute," Ohio University. www.biosci.ohiou.edu/qbi/

[Sus95] Susanne, C., (ed.) *Evaluation of Biology in the European Union*, VUB Press, Brussels, 1995.

[UW03] "Computer Science and Biology in the University of Washington Department of Computer Science." www.cs.washington.edu/building/bio.pdf.

[Turing01] Turing, A., "On Computable Numbers, With an Application to the Entscheidungsproblem," *Proceedings of the London Mathematical Society*, Series 2, Volume 42, 1936; reprinted in M. David (ed.), *The Undecidable*, Raven Press, Hewlett, NY, 1965. www.crumpled.com/cp/ classics/turing1.html.

[Whea03] "Genomics Research at Wheaton College." genomics.wheatoncollege.edu.

[Wick00] Wickware, P., "Next-generation biologists must straddle computation and biology," *Nature* 404 (2000) 683–684.

Building Connections in Research Universities

Wendy Katkin and Gayle Reznik
*Reinvention Center at the State
University of New York, Stony Brook*

Wendy Katkin is Director and Gayle Reznik Assistant to the Director of the Reinvention Center at the State University of New York, Stony Brook, which is devoted to improving undergraduate education at research universities.

Research universities are at the center of the post-genomic revolution in biological research, which increasingly relies on advances in mathematical and computational sciences. So it is natural to inquire into how, or if, these same institutions are modifying their undergraduate programs to reflect the increasing intersections of these disciplines in modern biology. This paper reports on a study focused on just such questions:

- How are undergraduate biology, computer science, and mathematics programs getting students to understand the complexity of modern biology and the critical role mathematical and computer sciences play in addressing this complexity?

- How do these programs bring together what is happening scientifically with what is happening educationally to foster change? How do they translate what is happening in research labs into the classroom?

An equally important interest was to examine curricular reform in light of the unique characteristics of research universities:

- How are individual universities approaching reform given the complexity of research universities and the reality that undergraduate biology and mathematics programs may lie in different or even multiple units?

- How are universities promoting change, given the constraints of

  — increasingly large numbers of students pursuing biology (including many with weak mathematical backgrounds);

  — large enrollments in most lower division courses, (which generally preclude individual, hands-on exercises; and

  — budgetary constraints (especially at public institutions) that prevent expansion of faculty?

This overview of efforts at research universities to strengthen connections in the undergraduate curriculum between the biological and quantitative sciences is derived from a study[1] conducted by the Reinvention Center. The goal of the study was to develop an information base that can be used to:

- Ascertain where undergraduate biology, mathematics and computer science at research universities are at the present moment with respect to the interface of biological and quantitative approaches;

- Inform both curricular changes and faculty development activities; and

[1] This study was supported by NSF grant No. DUE-0224652, Project Director: Wendy Katkin.

A Real Sea Change

When I started doing this, there was a lot of scorn among cell biologists and development biologists about the role mathematical models could play. I think that was justified scorn because there were too many early models that were fantasies and not attached to any biological reality.

There is a real sea change at NIH and NSF and among real biologists. All of a sudden they have so much real hard data that they can prove false virtually any mathematical model that somebody proposes. So, now there is real balance between data constraining models and perceived need for mathematical reasoning that is growing very fast. And that is a surprise to me, but it is a really welcome surprise.

Biology is so complicated, so much more so than, say, airplane design, that there is an extremely greater need for mathematics in biology than in Boeing to simulate their next plane. Many people are making really realistic mathematical models that are built from the ground up on real biological measurements. A very refreshing lesson is that biologists are now starting to take mathematics seriously, I think for the first time.... For a long time people took mathematical models seriously in a general way, but now real biologists are taking it seriously as shedding light on the phenomena they study.

- Provide a foundation for universities to move forward in efforts both to integrate quantitative approaches into the undergraduate biology curriculum and to incorporate knowledge of biological concepts and techniques into undergraduate curricula in the mathematical and computer sciences.

The study found a high level of interest and activity among participating institutions. Activities range from the development of new undergraduate courses and modification of existing ones to new faculty hirings. For the most part, however, these initiatives have been stimulated by factors not directly related to undergraduate education. These factors include research advances which give rise to the need for faculty with new skills and specializations; increasing research collaborations among biologists, mathematicians and computer scientists (95% of the participating institutions reported such collaborations); new computational technology; increasing interdisciplinarity in graduate programs; and, to a lesser degree, response to a "perceived need" in the education of future biologists.

Often one or two individuals led the effort. Thus many of the recent educational changes with a bio/math focus have been uncoordinated. Few of the departments involved in the study have undertaken a systematic examination of their undergraduate curriculum for the specific purposes of strengthening the connections between the biological and quantitative sciences. Similarly, few have taken steps to ensure that mechanisms are in place to help students develop both a conceptual understanding of the interconnections and sufficient quantitative literacy to pursue research that is at this intersection. This paper describes some of the most common patterns and trends.

Method

The study consisted of three activities: two surveys and follow-up interviews, all of faculty at research universities.

- A survey of directors of undergraduate biology designed to establish how their life science departments are approaching the intersection of biology and mathematics; to discover how they integrate concepts and techniques from the mathematical and physical sciences into undergraduate biology education; and to identify exemplary approaches which could serve as models for research universities.

- A parallel survey of directors of undergraduate studies in mathematics, applied mathematics, and computer science to determine whether their departments offer opportunities for undergraduate majors to learn about the role of mathematics in the biological sciences and to participate in research on biological topics.

- In-depth interviews with two groups: faculty in a biological, mathematical or computer science department who are collaborating at the research level, and faculty in these disciplines who have played leading roles in developing new undergraduate courses that address the connections between the biological and quantitative sciences or are designed to improve students' quantitative or biological literacy.

Survey of Directors of Undergraduate Studies of Biology

The survey on the Integration of Quantitative Approaches in Undergraduate Biology (hereafter

Biology Faculty Survey[2]) was a two-part instrument that allowed for multiple responses and, in some cases, open-ended comments.[3]

The first part was designed to elicit information about the overall curriculum, goals for students, areas of emphasis, and efforts to introduce quantitative approaches in the respondent's department or program. It also sought to identify factors motivating the reform efforts and factors that were deterrents.

The second part addressed issues specific to individual faculty such as their personal goals for undergraduate biology majors; the extent to which their department's current curriculum satisfied these goals; their knowledge of reports such as Bio 2010 that recommended strengthening quantitative skills for biology majors; their response to such reports; and the types of courses (i.e. introductory, laboratory, advanced) in which they would like to see a greater emphasis on mathematical, statistical, and computer science concepts and techniques.

The survey was sent via email to the directors of undergraduate biology at all 123 research universities (using the old Carnegie Foundation designation), along with a note explaining the purpose of the survey and inviting them to participate.[4] Twenty-two individuals (18%) responded. Although this number is low, the respondents represented a range of public and private research universities from different regions and there were sufficient commonalities among the responses to discern patterns and trends.

Survey of Directors of Undergraduate Studies of Mathematics, Applied Mathematics, and Computer Sciences

The survey on Educating Mathematics and Statistics Majors about Potential Applications of Quantitative Concepts and Techniques in Biology (hereafter Quantitative Faculty Survey)[5] was developed and imple-

mented in a similar fashion to the Biology Faculty Survey. This survey was designed to assess the extent to which undergraduate curricula in mathematics and computer science include mechanisms that encourage and prepare students to participate in research in a biological field, and also to determine the varied roles mathematics and computer science faculty play in efforts to incorporate mathematical concepts in undergraduate biology.

As with the Biology Faculty Survey, Part One focused on curricular and co-curricular issues while Part Two focused on issues specific to faculty. Twenty-four directors of mathematics and applied mathematics programs (19.5%) and 22 directors of computer science programs (18%) responded—a response rate consistent with the rate among the undergraduate directors in biology, and also representing a broad institutional range.

Interestingly, there was no overlap in the institutions of the respondents to the two surveys. Thus despite the low response rate to each individual survey, together the surveys provided information about integration efforts at 68 (55%) research universities, thereby giving credence to the broad findings. Their reliability was reinforced by the observations of those interviewed who generally substantiated and expanded upon the survey responses.

Faculty Interviews

Thirty-four faculty participated in interviews of approximately one hour.[6] Within this group, 21 were in a biological science, ten were in mathematics, two were in statistics, and one was in computer science. Thirteen (10 in a biological science and 3 in a quantitative science) of those interviewed were involved in the development or teaching of a course that emphasized connections between the biological and quantitative sciences or that was developed specifically to improve biology majors' quantitative skills. Twenty-one faculty (11 in biology and 10 in a mathematical science or computer science) were doing research at the intersection of biology and a quantitative science; about one-fourth of this group were teaching undergraduate courses related to their research interests.

The purpose of the interviews was to capture nuanced information not readily accessible through the surveys. Faculty who were collaborating in teaching, for example, were asked how their course originated and what its goals were, who participated in its design and teaching, whether their departments supported their efforts, in

---

[2] The Biology Faculty Survey, as well as the Quantitative Faculty Survey and the Faculty Interview Protocols can be found on the "Challenges" project web site at www.maa.org/mtc/ and on the Reinvention Center web site: www.sunysb.edu/reinventioncenter.

[3] This survey and the Quantitative Faculty Survey were developed by Dr. Katkin and Dr. Barbara Lovitts, a sociologist with considerable experience in survey and interview methods. The surveys were pilot tested by Dr. Lovitts.

[4] The Reinvention Center is grateful to the American Society for Microbiology (ASM), which distributed the survey and collected, collated and analyzed the data.

[5] An identical survey focusing on computer science education was sent to directors of undergraduate studies in Computer Science departments. For the purposes of this overview, the findings of the two surveys are combined.

---

[6] The interview formats were designed by Project Director Katkin and Dr. Lovitts and pilot tested and conducted by Dr. Lovitts. Gayle Reznik of the Reinvention Center analyzed the responses.

what majors were the students who typically took the course, and what were the institutional and disciplinary challenges they faced. Faculty who were collaborating in research were asked about the nature of their collaboration, the different skill sets and orientations they and their co-researchers bring to the effort, the issues they encounter as a result of their different disciplinary orientations, and the knowledge and skills they seek in undergraduates who want to work with them. They were also asked to consider whether their own research could be translated into an undergraduate course, to help students see the connections between biological and mathematical fields in addressing complex problems. All faculty who were interviewed were probed for their perspectives on the kind of preparation future researchers working in their area will need and the implications this has for undergraduate education in biology and mathematics.

Biology Faculty

The study revealed several common themes. Foremost, the quantitative concepts and models that are now required in much research in biology are finding their way into the undergraduate biology education offered at research universities through new and revised courses, individual student research experiences, and efforts by many faculty to describe their research to undergraduates and bring research-related problems and data to their classes. Further, strengthening students' quantitative skills so that they understand and can apply these concepts in a meaningful way is definitely on the agenda of virtually all biology departments. Although few campuses have undertaken a major examination or revamping of their basic biology curriculum, participants in the survey pointed to considerable activity that is taking place within the biological sciences to develop students' quantitative skills. Biology faculty appear satisfied with the pace of change in this area, as well as with their department's current curricular offerings and approaches to incorporating quantitative content. Figure 1 shows the extent these various approaches are being employed on respondents' campuses.

When asked how well their program educates majors in quantitative concepts and techniques, 68% of the directors of undergraduate studies responded "moderately well," and 14% responded "very well." Only 18% responded "minimally," and none responded "not at all." Seventeen faculty (77%) indicated that their departments had revised elements of the curriculum to strengthen the mathematical (18%), statistical (23%) and/or computational (36%) components.

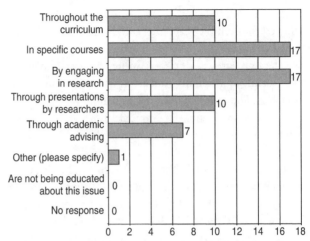

Figure 1. How are the students at your institution being educated to understand the complexity of modern biology and to recognize that they will need to understand quantitative concepts and methods *for further study and research?*

The two main approaches to change in introductory level courses have been to revise existing courses or labs (59%) or to create new courses within the department (32%). Four types of change relate to advanced courses and labs: revising existing courses (36%), creating new courses within the department (54%), creating new interdisciplinary courses (41%) and introducing new requirements for majors (46%).

Altogether, 82% of the biology programs that responded to this survey now offer courses that "address the intersection of biology and mathematics, statistics and/or computer science." Five participating universities established a major in computational biology or bioinformatics in the past five years, and during the same time period six universities established minors, also in computational biology or bioinformatics.

New Courses

The interviews provided insight into three types of new courses that have been created in recent years: (1) Introductory level mathematics courses for life sciences students that replace or revise the standard calculus sequence; (2) introductory level general biology or life science courses that incorporate more quantitative aspects; and (3) advanced courses at the intersection of biology, mathematics, and computer science that are directed at majors in all these disciplines. Introductory courses almost uniformly address the concern that biology students are not being adequately prepared in quantitative concepts and techniques to continue the study of biology.

New introductory-level mathematics courses have such titles as "Mathematics for the Life Sciences,"

"Techniques for Mathematical Biology," and "Applied Math in Biology." The revisions to the introductory life science courses generally incorporate additional quantitative materials or revise quantitative components of lab courses in order to give students experience working with graphs and computer software.

Advanced courses typically serve two functions: To further the quantitative training of biology majors and to introduce biological applications to mathematics and computer science majors. The largest number of new courses developed in the past five years are in the third group. The most common topics are biostatistics, bioinformatics, mathematical biology, computational biology, ecology, population biology, population genetics, mathematical biomechanics, and modeling of biological systems.

Sample Courses

We describe one example of each type of new course. The first two retain the traditional lecture format in order to accommodate the large number of prospective biology majors research universities typically teach.

Type One

Ten years ago, in response to a perception among both faculty and students that the standard calculus sequence was not serving the needs of biology students, faculty at the University of Toronto developed "Biology, Models, and Mathematics," a one-year introductory level mathematics course for biology students. A committee made up of faculty in biology and mathematics was formed to

consider alternative approaches that would help students understand essential mathematical concepts and see the relevance of these concepts to biology. After first considering parallel mathematics and biology courses that would cover coinciding topics, the committee decided instead on a single course in which biology would be integrated directly into the mathematics curriculum.

A mathematician and 2–3 biologists teach the course, which consists of lectures accompanied by weekly tutorials (recitation sections). In addition, students must also enroll in an Introductory Biology course as a co-requisite. The co-requisite was introduced in an effort to ensure that undergraduates who take the "Biology, Models, and Mathematics" course are genuinely interested in biology. The concern was that students in mathematics or computer science or engineering with no interest in biology might take it primarily because they perceived it as easy. Approximately 100 students enroll in the course per semester.

During the lecture component, a biology professor typically introduces a biological problem selected from the areas of population genetics, conservation biology, evolution, growth, population dynamics, physiology, or cell biology that requires certain mathematical tools to solve. The mathematics professor then educates the students about the theory surrounding the tool and teaches them how to use it and how to apply it specifically to the biological problem. The mathematics topics covered during the two semesters include linear regression, logarithms, power functions, logarithmic graph paper, exponential and logistic growth, elementary probability, derivatives, integration, dynamical programming, differential equations, Markov chains, and a brief introduction to chaos theory. Students can earn credit in biology for this course, but they may not earn credit in mathematics. (Contacts: Joe Repka, Department of Mathematics, and James Rising, Department of Zoology)

Type Two

At the University of Tennessee, discussions at an NSF-funded workshop held in the 1990s for biologists and mathematical biologists led to revisions in the 100-level "General Biology" course then being offered. The revisions revolved around the development of two types of instructional materials specifically for the course. The first is a set of 50 mathematical modules, all at the high school level, that are designed to correspond to topics in General Biology, namely either ecology and evolution or cell and molecular biology. The modules are all structured the same way. They start with an underlying biological question, move to a discussion of why that question is important, and then describe the appropriate mathematical approach to be used to answer the question. Typically, students are presented with a set of equations that represent the key variables and are taught how to measure the variables. They are then given data sets to analyze. At the conclusion of an exercise they are asked to respond to questions that require them to demonstrate some understanding of the mathematics that they did not understand prior to the exercise.

Since the lectures can have anywhere from 200 to 500 students and instructors vary from semester to semester, instructors are given great flexibility in determining how to apply the modules. Some incorporate small pieces of modules in lectures, others use individual modules as mini-components of the course, and others adapt them for homework assignments or extra-credit assignments. The use varies from instructor to instructor and from semester to semester.

The second type of instructional material developed for the "General Biology" course is a lab manual designed to be used in the lab sections that are offered in conjunction with the lectures. The lab manual focuses on quantitative methods, particularly statistical applications. (Contact: Louis Gross, Department of Ecology and Evolutionary Biology)

Type Three

A typical example of an advanced interdisciplinary course is the elective "Applied Mathematics in Biology"

that is team-taught by faculty in biology and mathematics at Utah State University. This course evolved from informal discussions between a biologist and a mathematician who were concerned about the lack of understanding among biology majors of basic mathematics concepts that are critical to study in a wide range of biological fields and a similar lack of knowledge among mathematics majors of modern biology, which increasingly relies on quantitative applications. The course has two components: lectures in which students study a series of mathematical topics and a corresponding wet laboratory in which they relate these topics to actual laboratory experiments. The course is listed by both the biology and mathematics departments and attracts students majoring in both. The students collaborate in both the lecture and laboratory study phases of the course. (Contacts: James Haefner, Department of Biology, and James Powell, Department of Mathematics and Statistics)

New Pedagogy

The recent curricular changes have been accompanied by significant changes in pedagogy as departments experiment with different approaches in order to engage the full range of students and promote the development of analytic and problem solving skills. The most common approaches incorporate inquiry and problem-based exercises into lectures, increase use of technology (often using graphics software), and give students hands-on experience addressing problems. Many of the experi-

The Only Way

I took an undergraduate statistics course and thought I knew statistics until I sat down in the lab where I was doing my undergraduate project. I tried to apply those stats to a fairly simple data set and I just realized I couldn't do it. I had to have my mentor come in and point me in the right direction, tell me which statistical analyses would be most appropriate, and why.

So I think it's a combination of coursework plus practical hands-on training of doing a project or a thesis. That is the only way. Students should take a well-taught course in elementary statistics and right after they should do an undergraduate thesis or project where they are going to have to think about the data they are generating and how they are going to apply math to analyze it or to test predictions.

mental practices such as hands-on emphasis, use of technology, and team work by students are designed to counter students' weak—"really atrocious," "pretty inadequate," "pretty poor"—quantitative backgrounds.

At the same time, some other trends such as increased involvement of undergraduate teaching assistants and creation of more, smaller labs and discussion sections are intended to address the large size of many classes.

While the dominant mode of instruction remains a sole teacher, faculty are working out ways to share the load. These include having professors with expertise in different areas—for example, a statistician, a computer scientist and a geneticist—teach different parts of a course. Another model is for a biologist and mathematician to attend all or most lectures of a course and teach in tandem. The biologist "presents an issue in biology" and then the mathematician "with those questions hanging in the air" "teaches the student the appropriate mathematical tool." "Once that's available, between the two of us we use the tool to consider the question and then draw some conclusions." A third model is the mathematics course in which the lecture is linked with multiple lab sections led by biologists in which the students apply the mathematics in the context of biological data and experiments. Figure 2 shows the range of pedagogical changes that have been made by responding departments in the past five years.

Support Issues

Together, these curricular and pedagogical changes show the attention departments are giving to undergraduate biology education and the level of experimentation. It is noteworthy that support for curricular innovations is coming largely from the biology programs themselves—either through internal funds or other local sources or by a department viewing the course as part of the instructor's regular teaching load. Only eight (36%) survey respondents and eight interviewees (24%) reported receiving external funds to develop new courses or introduce new pedagogies. One biology interviewee summarized the situation well when he said that new courses with a quantitative emphasis would have been developed in his department with or without external funds because of the "intellectual need for them." Another interviewee, who also noted that funding was not a problem, pointed to a second, equally important factor that was impeding further curricular and pedagogical advances: "It's not money that has kept us from going faster, although we are going pretty fast. It's [the need for] more time." Money however can help faculty "buy" the time.

The only situations in which funding seems to be a major factor is when a faculty member's department does not support the innovation—as when, for example,

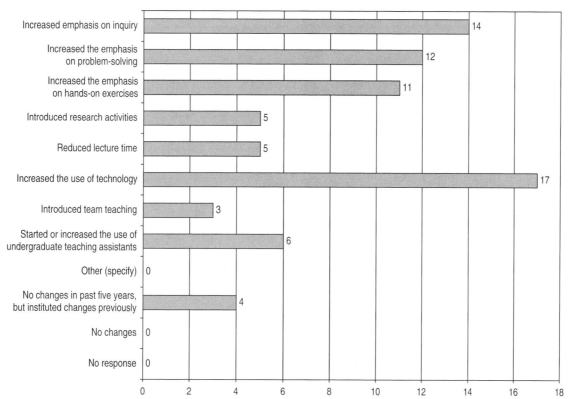

Figure 2. Which pedagogical changes has your department made in the past 5 years?

a biology faculty member taught mathematics to undergraduate biology majors even though "in the view of some of my colleagues biologists needed mathematics like a fish needs a bicycle." This problem, however, does not appear to be widespread within biology departments where, interviewees reported, colleagues generally appreciated their efforts because they made the department "feel more rounded," gave "mathematical confidence" to undergraduates and "launched" them into programs that involved mathematical modeling. All the interviewees indicated that their chairs were supportive of their course, some more and some less. Six characterized the chair as "pretty," "very" or "very, very" supportive of the course. Another professor, to show the extent of the support, pointed out that the division director and department head let him teach the course even though the department did not get credit for the course. At the same time, at many universities the question of funding lingers and may affect further progress: "We often had to teach them [courses] ourselves on and over and above hourly teaching load level because the university just did not have the resources or the manpower to hire new people."

Most of those interviewed who were involved in course development expressed pride in their course and felt that, although it was similar to courses at the graduate level, it was unique to their department at the undergraduate level. A typical comment: "Mine is probably

the only course at the undergraduate level that really goes into the quantitative aspects ... where I'm really trying to make sure they understand what they're doing. It's not just a black box where you do some modeling ... [but] you really don't understand how it comes out."

Most said that their courses fit in with their department's curricular reform efforts. They offered as examples their department's interest in increasing the quantitative education of students; the existence of a curriculum task force or committee that has as one of its goals improving students' quantitative training; other faculty in their department adopting for their own courses some of the problems and modules that had been developed for the course; the course being made a requirement for either the major or minor; and their department making mathematical biology part of its graduate program.

One biology professor criticized her institution for not having "any sort of major undergraduate curriculum review of mathematics as a whole, meaning mathematics, statistics and computer science" for many years. Another biology professor commented that his "department is committed to making sure our students are exposed to mathematics as much as we can" but it is difficult because of all the requirements students need to meet. This sentiment was echoed by several others. Only two of the interviewees felt that their courses did not fit in with their department's curricular reform efforts. A biology professor said that since his course is an elective, the department does not promote it very well. A mathematics professor pointed out that the course he developed is outside the mathematics department, but is instead an integral part of the biology department. He has modified the course in response to changes in the biology department's curriculum.

Quantitative Faculty

The situation within mathematics and computer science is much more complex, largely because educating their undergraduates to participate in biological research is only one of several goals these departments may have for their students and it is not central to their missions. Moreover, biology is a relatively new interest for these departments. This is especially true for computer science departments, which, for the most part, have undertaken undergraduate initiatives with this focus only in the past five years. Nevertheless, there is much more activity in this area than one might expect.

Much of the activity derives from the high level of collaboration in biology-related research between faculty in these departments and colleagues in biology, bio-

Mathematics as a "Fraternity Initiation"

In my judgment, mathematics departments have done an absolutely terrible job teaching mathematics to undergraduates at almost every university everywhere, and I am including myself in that; I used to be in a mathematics department. Mathematics is taught in the most peculiar way that offends most normal people rather than draws them in ...

In almost all of the natural sciences the early courses are survey courses that show you the wonders of the subject, without bogging you down in details. They entice you to pursue biology because you get a glimpse of all this amazing phenomena that you can learn about.

In contrast, mathematicians teach mathematics in some kind of step-by-step process, almost like a fraternity initiation rite. "Because I did it this way, you students are going to learn first to differentiate 'e to the 1 over arctangent of u squared' with a pencil." And mathematics curricula teach univariate calculus before introducing multivariate calculus. There are no interesting phenomena that you can tackle with univariate calculus. Students end up not even knowing what a derivative is after they've taken mathematics for a long time.... They only know something about the definition of slope of the tangent line.

It just doesn't make sense the way mathematics is taught in this brick-by-brick way, where if you stick with it for six years you'll finally understand what the point was.

I wanted to figure out how to teach a course where I told students right at the beginning what the point was, what a derivative is: it is a linear map that is tangent to a given nonlinear map at a certain point. Never mind that it happens to be the slope of the tangent line if you are talking about a single function of one variable. I wanted to teach students what dynamical systems did, and teach them about the behavior of nonlinear dynamical systems before getting bogged down in the details of linear ODE systems that you need to understand the local behavior in your critical points.

It took a while to put together an approach that would do this, that would really entice students and not offend them, that would show them the big picture before they got bogged down in details. I now see, in retrospect, that mathematics should be taught like a Berlitz language course—because it is a language and not a science. You should just start speaking this language the way mathematicians really speak it until by osmosis the students start to become fluent in the language. All this comes before they have to do all the detailed epsilon-delta stuff. It took me a while to figure out how to create a course that had some real substance but that was light on the details of formula manipulation that seem to offend so many students and drive them out of mathematics.

chemistry, bioengineering, units within medical schools, units within schools of agriculture, and interdisciplinary research centers on campus: 71% of the mathematics and applied mathematics departments and 91% of the computer science departments that participated in the survey reported having such collaborations. Undergraduates often participate in the collaborative projects, which in some cases appear to be the main research option available to them. In 19 mathematics departments responding to the survey, for example, in which fewer than 10% of the undergraduate majors participate in research, *all* are involved in biologically-related projects. Among 20 responding computer science departments in which up to 25% of the undergraduates reportedly are involved in research, 86% work on biologically-related questions.

The array of subjects that these undergraduates are pursuing reflects the centrality of quantitative approaches to virtually all branches of biological investigation, ranging from genetics and microbiology to ecology and evolutionary and population biology to organismic biology to

A Common Obstacle

The usual obstacle to doing a mathematics course that is designed for a particular clientele is the math department's general unwillingness to offer specialized courses. The argument is always, "What happens in the course?" "Who do we get to teach them?"... It is a concern I hear regularly from math departments at other places. "We could never offer this course because it focuses on biological examples and we don't have the faculty to teach it."

pharmacology. Student research topics include, for example, statistical analyses of DNA sequences, brain pathology modeling, male/female mortality rates in different species, computational phylogenetics, cell signaling theory, population dynamics, mathematical modeling of carcinogenesis, development and analysis of models for the spread of an infectious disease, and ecological modeling related to spatial aspects of animal behavior.

The interviews suggest that mathematicians, computer scientists and even biologists who collaborate in research prefer working with undergraduates with strong quantitative skills precisely for the skills they bring. Some typical comments:

- "[Biology students would need] whole courses before they are going to be useful for programming purposes."

- "When they get to the data analysis part of the project, that is when most undergraduate students fall flat on their faces. Math is more of a skill, it is just as much of a skill as it is an intellectual concept, and so applying math is a skill that has it be taught."

- "It's relative easy … to train the computer students to give them enough information about biology that they can be useful. That is pretty easy, but it is almost impossible to go the other way in a short period of time…. Students from the life sciences are not going to be useful at all for programming part of these projects unless they've already had significant programming experience."

Interdepartmental collaboration at the undergraduate level, however, is not as widespread. Only 8 (33%) mathematics and applied mathematics departments and 8 (36%) computer science departments reported having such collaborations. Moreover, many of these interactions revolve around specific programs or course groupings, such as bioinformatics and computational biology. At the same time, half of the mathematics departments and 41% of the computer science departments surveyed offer their own courses that include, to varying degrees, segments intended to "stimulate students' interest in applying quantitative concepts and techniques to biologically-related courses" and to "make undergraduates aware of opportunities to engage in biologically-related research." While some of these courses have a specific topical focus, especially those more recently developed, the majority are designed to give students general background in biology or experience applying their knowledge and skills in a biological field.

The number of specialized courses and interdisciplinary programs is likely to rise significantly in the next few years as more and more faculty who are collaborating in research begin to translate their research into undergraduate courses. With three exceptions, all the "research" faculty who were interviewed indicated that they could envision new courses emerging from their research. Several noted that this is already happening, especially in recently-created and "topics" courses, and it promises to continue.

New Courses

The level of participation by biology faculty in the design of new mathematics, statistics, applied mathematics and computer science courses varies from minimal to significant and is perhaps impeded by a tension that often exists between biology and mathematics and computer science departments on how to teach the course. The tension arises over how much life science and quantitative content should be in the courses: "Disciplinary issues were driving what the focus of the course should be."

Another issue is the different way biologists, mathematicians, and computer scientists approach a course: Biologists "think in a looser way, perhaps because the world they deal with is looser," while mathematicians and computer scientists "think in a rather formulistic way," which affects their pedagogy.

Faculty who have collaborated in developing these courses recommend that future collaborators recognize from the outset their own differences in perspectives and the differences that exist between teaching mathematics majors and biology majors: "Unlike mathematics students, biology students are not as interested in the math itself and the elegance of math. Rather, they want to know how to use the math. Use biological examples and be more visual and describe equations in words when teaching biology majors."

Biology students "don't really like to take math classes, so if you can give them some biology to do and some mathematics to do, then it's a much better experience." "You want to keep them excited about the biology while they are learning the math." One biologist said that in order for these courses to be effective, planners need to "make sure the biological aspects and the mathematical aspects are fully integrated." Similarly, a mathematician observed, "one should care about both disciplines and think of both as disciplines."

The study also revealed differences in how colleagues from the various departments feel about the new math/bio courses. Biology faculty who participated in the survey and interviews almost uniformly thought

The Challenges of Interdisciplinarity

as Articulated by Two Mathematicians and a Biologist

Mathematician: Initially we spoke different languages so we had to learn to communicate among ourselves, let alone with the students. We all feel that we've learned a great deal from the experience. When I began this, I actually knew very little biology, having never taken a university course in it. In truth, I got involved in the first place partly because I thought it was an excellent opportunity to learn some biology.

In fact, it worked so well that within a couple of years I began a fairly substantial research collaboration with a biologist—to a large extent on the strength of what I am doing in this course. Initially it took us a lot of time. We had a lot of meetings and a lot of discussions, and it took quite a while to cobble together the course we eventually taught. It has continued to evolve, although it hasn't changed radically.

Our differences, ultimately never serious, were interesting. For instance, as a mathematician, if I needed something I'd just make it up, whereas the biologists are rather horrified by that. They are concerned with the sanctity of real data, from actual experience, from actual observations of real life. This wasn't serious, but it was interesting for me to see. If I could make up some numbers that illustrated my point, it didn't really matter whether they were real or not. But to biologists it was quite different. And so I would make up problem sets that included questions about imaginary plants living on Mars and things like that. My partners, I think, found this a little peculiar.

Similarly, I would get a bit startled by the lengths they would go to find actual data, when it seemed to me it would have been perfectly easy to make some up. That was an interesting cultural difference, but it was never a problem. In fact, it was quite interesting for all of us to see this difference and it was a source of some amusement.... There was a very good interaction among us and we were always able to find appropriate ways of resolving different perspectives. So, we really didn't have problems. We had some interesting experiences, but not problems.

• • •

Biologist: A major challenge is understanding what's important and what's not important in the field ... Scientists often have a very good instinct for what's important within their own fields, but that instinct usually severs completely when you are new to a field. Re-establishing instinctive feelings for what is going to be important and what isn't takes some time.

We also encounter faculty issues when somebody is between fields, e.g., how they judge, are they jealous. Say it's about mathematicians. Are they judged by biologists or are they judged by others? Do the biologists think they are really good mathematicians and do mathematicians think they're really good biologists? Or do we really want the opposite? These are real challenges and I think that we're going to see a lot more of them as time goes on because we're sitting in an area where there haven't been that many cross appointments.

Now we're becoming very aware of appointments that have people in both fields, and how exactly to handle them. I think it is going to be challenging. When it comes to bettering the academics, people in the computer science committee, for instance, will have a very hard time knowing why this work is important. Even if this is an important application of biological relevance giving forth new knowledge, the computer science might not be spectacular. Equally well, there may be a piece of work which involves some very nice computer science or mathematics, but people on the biology committee won't have a chance of evaluating it. Remember, you are evaluated by your peers ... I think that this is a problem that is going to have to be solved.

• • •

Mathematician: Interdisciplinary work ... is harder and less efficient than disciplinary work. That's because it requires more interaction. It requires taking viewpoints that are a little bit different and synchronizing them, if you will, to be able to bring them to bear on a problem. It takes time to learn the problem because you have interdisciplinary things going on. There may be difficulty in getting things published: if you send something with too much math to a biology journal, they may complain; or vice versa, if it's got too much biology and you send it to a math journal, then they complain.

So in various ways it's harder to do interdisciplinary work. It takes more time and energy than just doing straight mathematical research. And in a sense it's less efficient. More time is spent spinning wheels or going down blind alleys or just trying to find out basic background things or making appointments to consult with whomever you need to consult with. On the other hand, I think that it can take you to places you wouldn't get to just by staying within your discipline. I think it the interdisciplinary aspect makes the work more worthwhile at a given level of mathematical sophistication. So, if I'm using the same mathematical tools of roughly the same power, the results are more interesting if it's a situation where it ties better with biology. That's the up side

The Struggle to Integrate

The main issue is about working hard to integrate the two disciplines, to avoid teaching either of them in a vacuum, but teach them together. In our case we did it with people from different departments teaching different parts, but it could conceivably be done by only a single person if that person is knowledgeable enough. But whatever, it has to be done in a way that the parts really do work together, rather than just go off in their separate directions

• • •

Mathematician: Our course was an immediate success, at least as judged by students' responses. There was a brief attempt to make an analogous course in statistics and biology. (Statistics here is a separate department.) That one was less successful.

Interviewer: Do you know why?

Mathematician: I think it's a question of the individuals involved. I don't really know, but from a distance, it appears as if the statistician involved was less willing or able to incorporate the biological information in a thoroughly integrated way. He basically just taught statistics. And the biologist involved, perhaps, had difficulties communicating clearly the ideas in a way that would be conducive to the students making the connections for themselves.

I think in the course that I am involved with we've been very careful and successful at linking the two parts.... In this other case, they were less successful.... It was more lectures on statistics and some lectures on biology, and it was up to the students to figure out how they might be related. I think that is what happened, anyway. It ran for a few years and then disappeared.

these courses are a good idea and are generally supportive. In contrast, faculty in mathematics and computer science are more skeptical because the courses do not directly help develop mathematicians or computer scientists and were therefore peripheral to their own academic interests. Within this latter group, applied mathematicians and statisticians were generally more positive because of the benefits they could see students deriving from such courses, particularly in preparing for new fields and research opportunities.

Virtually every faculty member in these departments agreed, however, on the need to improve biology students' quantitative background. "Biology students tend to have weak mathematical training and the requirements allow them to remain this way." "The impetus has not been forthcoming to strengthen students' mathematical background." A frequently expressed alternative to the development and revision of courses is for biology departments to increase or change their quantitative requirements.

Some mathematics and computer science faculty who have had experience with integrative courses have become more positive: "[Biology students] learned their mathematics in a rote sort of way, and they really have no appreciation of what it's for. One of the good things about this course is that it actually teaches them to use some things that they've learned as rote skills. It actually brings [mathematics skills] to life for them."

A concern voiced by biology faculty is that faced with a choice between a biology course that is mathematically-driven and one that is not, many biology and pre-med students would choose the non-mathematical option. Another concern relates to funding, which appears to be a greater factor in impeding progress in mathematics and computer science departments than it was for the biologists. This comment from a mathematician is typical: "The effects of the economy are stressing our abilities to offer our basic courses—that are absolutely necessary. So when we consider offering a new course, we must do so carefully."

Sample Courses

The new mathematics, statistics, applied mathematics and computer science courses that have been developed with an eye toward biology can be grouped into three categories: (1) introductory level courses that revise or replace standard introductory calculus or statistics courses; (2) advanced electives, offered by these departments; and (3) interdisciplinary or jointly offered courses. While the initial electives and interdisciplinary courses were in such fields as bioinformatics, computational biology, and modeling, the range of subjects is increasing as more courses are being developed in more varied fields. The introductory courses typically target prospective biology majors. The departmental and interdisciplinary electives are usually geared for biology, mathematics, applied mathematics and computer science majors, as well as majors in related disciplines, such as bioengineering.

The first and second courses described here were developed at large public universities. The third course was developed at a large private university. The two upper division courses, it should be noted, target both advanced undergraduate students and graduate students—a common practice reflective of the small enrolment these courses typically have and the shortage of faculty available to teach them.

Type One

The two-semester sequence "Calculus with Biological Emphasis," offered at the University of Minnesota, is a variant of the standard introductory calculus course. The course was conceived and designed by Claudia Neuhauser, in response to concern among biology faculty that the standard calculus course was not giving prospective biology majors the kind of quantitative background they would need for further study and research in biology. At the time, Neuhauser, a mathematical biologist, was in the Mathematics Department; she has since moved to the Life Sciences. The sequence is taught entirely by faculty in mathematics. Although this variant covers many of the same topics as the regular first-year calculus course, it also includes differential and difference equations, matrix models, and some probability and statistics. It is distinguished by its emphasis on applications in the biological sciences and its use of word problems taken from biological research papers. The sequence is taught in a traditional lecture/recitation section format. The text, *Calculus for Biology and Medicine* (Prentice Hall, 2003), was written especially for such calculus courses that have a biological emphasis. (Contact: Claudia Neuhauser, Department of Ecology, Evolution and Behavior)

Type Two

The "Mathematical Biology" course at the University of Utah exemplifies an advanced course offered by a mathematical sciences department for its advanced mathematics majors and first-year graduate students. The course is designed to introduce students with a strong mathematical background but not necessarily any knowledge of biology to some of the basic models and methods of mathematical biology that are not usually seen in standard mathematics courses. The course covers models of population dynamics, reaction kinetics, diseases, and cells that can be written as ordinary differential ques-

tions, delay-differential equations, and discrete-time dynamical systems. (Contact: Frederick Adler, Department of Mathematics)

Type Three

In contrast to the previous example of a course developed and taught within a department of mathematics, the "Computational Biology" course at the University of Pennsylvania is offered jointly by the departments of Computer Science and Biology and is team-taught by faculty from both departments. Topics include computational problems in molecular biology such as sequence search and analysis, informatics, genetic mapping and optimization. The course was developed as part of a training grant for graduate students in computational biology, and thus is officially a graduate level course that advanced undergraduates are permitted to take. It attracts majors in biology, computer science, and statistics. Biology undergraduates who choose to take the course are typically concentrating in mathematical biology or computational biology. (Contact: Warren Ewens, Department of Biology.)

Although courses with a biology component are the primary means through which mathematics, statistics and computer science majors learn about applications of quantitative methods in biological research, students also learn about these connections through undergraduate research experiences and research presentations by faculty, graduate students and more advanced undergraduates. Survey respondents indicate that these non-course means are almost equally important. Thus the environment itself, with its strong emphasis on research, plays an educational role—particularly in acquainting students with the range of research projects on which they might work. Twenty-three survey respondents pointed to notices on bulletin boards and announcements made by faculty in class as means by which undergraduates also learn about research opportunities. Several respondents noted that the involvement of undergraduates in research in these areas was to a great extent spurred by the availability of funds through the NSF (e.g., VIGRE, Integrative Graduate Education Training [IGERT], REU supplements, and site grants), NIH, NASA and the Howard Hughes Medical Institute, as well as through major grants awarded to many of their faculty supervisors.

Conclusions

Undergraduate education in biology, mathematics, applied mathematics and computer science is changing

From Obstacles to Opportunities

From the biological side of things, about ten years ago our colleagues didn't believe that mathematics was important. ...They viewed it as a hurdle their students have to get over so that then they can do real biology.

I think that has changed completely in the last decade, at least in the U.S. Now you pick up any issue of *Science* and any of my colleagues will acknowledge that if you look at the jobs, they're quantitative. There's computational biology, there's genomics. There are all these areas that require very, very good quantitative skills. That is where the market is. They will not make that ten-year-old argument anymore.

Virtually all biologists now agree that strong quantitative skills are important to be able to read the literature and to be able to get to the point of doing thesis research ... The whole attitude towards mathematics changed because of the growth of genomics and computational biology. These present tremendous opportunities that have been pointed out in reports such as Bio 2010.

A variety of new groups have been formed within the National Science Foundation, the National Institutes of Health, and the Howard Hughes Medical Institute. These groups realized that future research requires a more interdisciplinary approach to biology, by which I don't just mean mathematics. Appropriate training, skill development, and conceptual understanding in key physical and chemical sciences components are really important. Along with these needs is the ability of different scientists to talk to each other. This means developing a language so that undergraduates can at least have some kind of coherent conversation with people who have somewhat different skill sets.

To me, that is the essence of appropriate undergraduate interdisciplinary training. It's not necessarily being able to have all of those skills yourself, but rather having the conceptual foundation upon which you can have an intelligent conversation with someone with a different skill set.

in response to several forces: the increasing complexity, changing direction, and new requirements of biological investigation in the post-genomic age; the need for future researchers and other professionals who have a combination of scientific knowledge and quantitative abilities; and, consequently, the changing composition of the departments themselves. The changes are evidenced in recently established courses and programs that are at the nexus of biology and the quantitative sciences, in the increase in joint offerings (though still small), and in the increasing interdepartmental collaboration of faculty, graduate students and undergraduates in research.

The extent and level of change is perhaps best evidenced in recent patterns in faculty hiring by all the departments that were part of the study. For all, the priority has been to bring in new faculty whose training and/or current work is at the cutting edge in these areas. Biology departments in recent years almost uniformly have been "proactively trying to hire mathematically skilled people," most commonly bioinformatics specialists, quantitative biologists, and computational biologists.

Fifteen of the responding departments (66%) hired one or more individuals with strong quantitative backgrounds. Within this group, seven departments hired three faculty for the specific purpose of strengthening connections between biology and quantitative departments; this number is particularly significant given the budgetary constraints under which most (especially public) research universities have been operating in recent years.

Similarly, 60% of the mathematics departments and 50% of the computer science departments have made "strengthening the connections between mathematics/ statistics [or computer science] and biology disciplines" the focus of their recent hirings and are looking for mathematical biologists, computational biologists, and biostatisticians. Two problems they have faced are a shortage of individuals with the required skills, as evidenced by four unsuccessful searches that were reported, and the high salaries demanded by individuals with the prerequisite skills.

While most departments move in this interdisciplinary direction with future research opportunities in mind, these new faculty will inevitably influence departmental thinking about the undergraduate curriculum as they provide advice and develop undergraduate courses that reflect their research interests. Judging by the interviews, this already appears to be happening. Four biologists and two mathematicians observed that they or their departments have taught courses which included their research focus, one recent hiree said he had been recruited specifically to develop and teach courses relating to his research and 18 faculty (nine mathematicians and nine biologists) were able to envision their research being translated into undergraduate courses.

Almost all the individuals interviewed who were involved in a research collaboration and also taught undergraduates indicated that they have incorporated elements of their collaborative research into their undergraduate instruction. They do so for example by bringing papers they have written to class, by using experimental data, problems and models derived from their research and by introducing current work and ideas of other researchers working on similar questions. In this way, they assert, students learn about ongoing work, see the relevance of specific materials, and gain an appreciation of the importance of thinking broadly about a problem, about the value of an interdisciplinary perspective. They also learn that there may be more than one correct approach and solution to a problem. The newly-hired faculty will also increase the department's capacity to offer more and a wider range of courses in this area.

The pressure to modify the curriculum appears much greater on biology departments than on mathematics or computer science departments because of the centrality of quantitative approaches to modern biology.

Equally important to biology faculty is a weakness they perceive in biology students' reasoning and quantitative skills. Many of the mathematics and computer science interviewees, for example, noted that even those science students who have taken mathematics courses "don't have an adequate conceptual basis for solving real world problems using mathematics." A biology professor aptly summarized the problem: "I would say a substantial portion of them choose biology for a major, as opposed to chemistry or physics, because the mathematics involved in either a chemistry major or a physics major frightens them and they view biology as a lesser of those evils. Students who have difficulty in quantitative abstract thinking are selecting themselves,… when the reality is that biology can become quite quantitative and has a lot of abstract concepts."

The changes in curricula that are occurring are evolutionary rather than systematic. They are stimulated primarily by such factors as research collaborations, personal interests of faculty, increasing presence of faculty with interdisciplinary specializations, and grant funds that support the development of new initiatives. They are

The Rewards: Two Perspectives

Biologist: There are always tradeoffs involved among research, administration and teaching. Obviously this course takes time and it takes time that I could be doing other things. So, I imagine in that sense it's negatively affected my research….

But it certainly has changed the way I've looked at things, and how I deal with mathematics. From time to time, when I have a mathematical question, because I know Dr. XXX well, I have no hesitation asking him how to do something. I've also been able to direct other people in my department to go and talk with him. On one case it's resulted in a collaboration between a zoology professor and Dr. XXX that's resulted in publication of several papers, and even co-supervision of some graduate students.

Mathematician: I feel that I've learned a lot about biology. I didn't take biology earlier because I hated having to memorize things, and I had the impression that the biology courses I would have taken involved a lot of that. So in a way I was happy to be able to short circuit the process and jump into the middle of it and have biologists around who knew the details so that I didn't have to memorize them and who could, of course, explain things to me more efficiently than I could learn them for myself.

So I had a crash course in biology and I believe that I really have learned a great deal of it in the intervening time and certainly enjoy it. It's a part of the scientific landscape of incredible importance; it's just good for me to know about it.

I have begun a quite substantial research activity with the biologists that wouldn't have happened if I hadn't been working on this course. It is an ongoing thing: we published several papers and are working on several more. So I feel that I have benefited a great deal, and I've also enjoyed it enormously. I don't know that I can speak at all for the others—the biologists—but I think they probably have come to feel more comfortable with some of the mathematical things.

Listen and Learn

You must make sure that a course like this is transferable to other instructors. I developed the course, I taught it for a few years, and then I moved into biology. I was still prepared to teach the course, but now because of the department that I'm in I can't teach that course anymore.

A course has to be developed having in mind that somebody else will ultimately teach it. So it cannot be something that is very time intensive, especially in large research schools where instructors very often just walk into classrooms and teach and don't expect to do much else.

It has to be user-friendly as well. It still has to provide the material to get the students excited. Not every professor at a big research university will spend hours developing and teaching a course. Most of them just want to walk in, prepare a little bit, and then teach. And you have to train your teaching assistants; you can't expect the average math teaching assistant to just pick up the ideas.

• • •

Be prepared for it to require a lot of work initially. This is probably the main advice. Secondly, do make sure that you have the administrative background all straight, that you clearly have the support of the departments, that they value this course as part of their program, and that they will encourage their students to take it....

Make sure you've clarified any issues about resources. It seems to me that a fairly common scenario in universities these days is that administrators are willing to allow faculty members to organize something but then at the last minute don't really come through with the funding, or cut it in half or something.... I've heard a lot of stories of people getting burned by having promises of support whither away before they were needed.

So I think it's important to try and pin that sort of thing down from the outset. Figure out what is needed and be clear about where it's coming from. I have a number of very frustrated colleagues who have tried initiatives (not this particular kind, but other things) and put a lot of time into it and then have the powers-that-be say "Fine, you can keep doing it, but we are not paying for whatever we were going to." That is very annoying and frustrating and leads to great dissatisfaction on the part of the poor sucker who got stuck this way. So I think it is crucial to make sure that it is all settled.

also being driven by the common recognition—articulated by every individual interviewed for this study—that it is "important," or "very important," or even "absolutely essential" (as one biologist stated) that undergraduates gain interdisciplinary skills and perspective as preparation for graduate study.

The survey responses and interviews suggest several persistent challenges that prevent a more comprehensive approach: existing demands on individual departments which in some instances are struggling to cover their basic courses; lack of funds for new positions "to hire the right people;" time constraints; different orientations and level of interest among faculty in different disciplines; and few incentives (for mathematics and computer science faculty in particular) to develop new courses directed primarily at undergraduates in biology. Several individuals also cited administrative problems such as determining which department receives credit for a jointly taught course and figuring out how new courses with blended emphases related to other departmental offerings. The most pervasive challenge however relates to

how interdisciplinary courses of this type should be conceived and what the goals and content should be.

Whether in biology or a quantitative science, based on their experiences, the faculty who participated in the study and were involved in creating new courses offer advice to colleagues who plan to develop similar integrative courses. Some of their recommendations are particularly relevant to faculty at research universities since they relate to such factors as the large class size at these institutions, and the likelihood that over time the instructors will vary.

Among the most common pieces of advice are these:

- Make sure the new course is user friendly and transferable.

- Only faculty with tenure should undertake such efforts.

- Expect to do a lot of work, clarify all the critical issues at the outset, and make sure you have the necessary administrative support.

Despite the challenges, there was consensus that curricular efforts to connect biological and quantitative sciences would continue and expand, especially as faculty begin to recognize the benefits to themselves as well as their students. Some of the benefits cited were the formation of networks; new opportunities to collaborate in research and teaching with colleagues from other departments; a new, reinvigorating way of looking at one's own research; a change in research focus or even a new focus; and gaining an appreciation of other disciplines and of how other disciplines approach problems.

A biologist summarized well the primary benefit faculty derive from developing and teaching integrative courses. It is this benefit that ensures that integrative efforts will persist: "Teaching is the best way to prepare you to get into a new field, particularly if you're talking about something on the interface between mathematics and biology."

Quantitative Initiatives in College Biology

Profiles of Projects at Undergraduate Institutions

Debra Hydorn
University of Mary Washington

Stokes Baker
University of Detroit Mercy

Jeffe Boats
University of Detroit Mercy

The following profiles of campus initiatives in quantitative aspects of undergraduate biology is the result of a targeted study of interdisciplinary efforts at two-year and four-year colleges and comprehensive universities undertaken by Project Kaleidoscope (PKAL) at the Independent Colleges Office (ICO) in Washington, DC. The goal of this effort is to inform those involved with undergraduate education in both the biological and mathematical sciences of the current state of integration between the two fields. The PKAL effort was lead by a project team of faculty who are part of PKAL's Faculty of the 21st Century (F21) initiative: Debra Hydorn, Chair of Mathematics at the University of Mary Washington; Stokes Baker, Associate Professor of Biology at the University of Detroit Mercy; and Jeffe Boats, Assistant Professor of Mathematics, also at the University of Detroit Mercy. After surveying the community to identify exemplary programs that translate collaborations into the undergraduate curriculum, they interviewed faculty leaders to profile how students at different campuses are being prepared for the "New Biology."

Capital University	Pomona College
College of Wooster	Seattle Central Comm. College
Davidson College	University of Detroit Mercy
Duquesne University	University of Mary Washington
Harvey Mudd College	University of Redlands
Hope College	University of Richmond
James Madison University	Univ. of Wisconsin-La Crosse
Macalester College	Wheaton College
Montana State University	

Computational Sciences Across the Curriculum

Capital University

Located in a suburb of Columbus, Ohio, Capital University is one of the largest Lutheran-affiliated universities in the country. Half of its 4000 students are undergraduates pursuing one of six different bachelor's degrees awarded by the university. Capital offers one of the nation's few undergraduate programs in Computational Science Across the Curriculum (CSAC).

Computational science is an emerging and rapidly growing interdisciplinary field at the intersection of

mathematics, computing, and science that investigates ways of using mathematics, computing and visualization to solve complex scientific problems. Capital University mathematician Ignatios Vakalis notes that many consider computational science to be a "third methodology in the development of scientific knowledge, alongside theory and experimentation."

Funded by grants from the National Science Foundation[1] and Battelle, Capital's undergraduate program in computational science is designed primarily for mathematics, science, and pre-engineering majors. The two chief goals of CSAC are to present mathematics within the context of science problems and to demonstrate how computing technology (symbolic, numeric, parallel, graphical/visualization) is used to solve problems from different scientific disciplines. Since modeling is an integral part of any computational experiment, all courses have a mathematical component.

Currently the CSAC program is offered as a minor that complements any mathematics or science major. For example biology or mathematics majors can supplement their major with the formal minor in computational science. All computational science courses make use of inquiry-based pedagogy, and all homework and projects are team based, pairing a science major with a mathematics or computer science major to work together on a single project. All projects require full written reports and oral presentations.

A number of computational science courses have been modified to include biology applications, and some have been developed specifically for computational biology. For example, one of the modules in Computational Science I[2] is the investigation, creation, and solution of mathematical models for the spread of infectious diseases. Students are guided to develop an SIR (Susceptible, Infected, Recovered) model[3], generate the appropriate non-linear coupled system of differential equations, use Maple[4] to solve the system numerically, use graphing tools to visualize the solution, and assess the results. Multiple homework projects are assigned for investigating extensions of the SIR model. In addition, case studies for modeling the spread of Malaria and SARS are presented using the STELLA[5] modeling package.

The CSAC curriculum[6] consists of two courses in Computational Science, one (Computational and Applied Mathematics) in differential equations and dynamical systems and a capstone team-based undergraduate research experience. Elective courses include two in parallel and high performance computing; one in scientific visualization; and an array of courses on computational methods in biology,[7] chemistry, environmental science, finance, psychology, neuroscience, and physics.

The impact of biology on CSAC is already broad, and still expanding. In addition to Computational Science I, the courses in parallel computing and computational mathematics have been enriched with modules that address the integration of biology concepts with mathematical modeling and computer implementation. In addition, the mainstream calculus courses include more applications to biology, and a new calculus course geared specifically to the life sciences is in development.

Modules and other educational materials that are generated for those courses address real-world issues through the following paradigm:

$$\text{Problem} \to \text{Model} \to \text{Method}$$
$$\to \text{Implementation} \to \text{Assessment}$$

All are taught using inquiry-based methods and involve students in interdisciplinary teams. Many students conduct research in bioinformatics with companies at the Columbus Business and Technology Center, the Ohio Supercomputer Center, and the Medical Informatics Division at Ohio State University.

In addition, with funding from the W.M. Keck Foundation, Capital University leads the twelve-member Keck Undergraduate Consortium for Computational Science Education (KUCSEC).[8] The goal of the KUCSE project, and its 26 faculty co-PIs, is the creation of online educational materials for teaching a variety of undergraduate computational science courses.[9] All materials demonstrate the connection of mathematics with a variety of disciplines. Examples from biology include modeling tumor-immune interactions,[10] modeling malaria,[11] and gene identification.[12] Capital plans to develop an additional course in bioinformatics and will soon offer a

[1] NSF CLLI-EMD grant (DUE: 9952806)

[2] www.capital.edu/acad/as/csac/Comp_Sci1/compsci1.htm

[3] mathworld.wolfram.com/Kermack-McKendrickModel.html
www.math.duke.edu/education/ccp/materials/diffcalc/sir/sir2.html

[4] www.maplesoft.com/products/maple/

[5] www.iseesystems.com

[6] www.capital.edu/acad/as/csac

[7] www.capital.edu/acad/as/csac/Comp_Bio/compbio.htm

[8] www.capital.edu/acad/as/csac/Keck

[9] www.capital.edu/acad/as/csac/Keck/modules.html

[10] www.capital.edu/acad/as/csac/Keck/modules/Depillis/depillis.html

[11] www.capital.edu/acad/as/csac/Keck/modules.html#malaria

[12] www.capital.edu/acad/as/csac/Keck/modules.html#gene

certificate program in Computational Biology/ Bioinformatics via its Summer Science Institute.

Contact: Ignatios Vakalis < ivakalis @ capital.edu>

Developing Computer Information Skills

College of Wooster

The College of Wooster is a small liberal arts college located in the rural community of Wooster, Ohio with an enrollment of 1700. The institution is strong in the humanities, and has approximately 20% of its students majoring in science and math. Biology is the largest science major, with approximately 80 students.

The availability of web-based resources makes genomic information and analysis tools readily available to undergraduates. Thus, institutions of all sizes and resources can train their students in the use of bioinformatics tools and in connecting these skills to the primary scientific literature. Faculty at the College of Wooster have taken advantage of these resources to reform their Biochemistry and Molecular Biology program. One of the driving forces behind their reform efforts is the National Research Council's Bio2010 report. Although Wooster faculty would like to add a course in bioinformatics, with only seven faculty teaching in the life sciences there is no room to add such a course. Adding bioinformatics-based assignments to existing courses represents a pragmatic solution to this problem. Additionally, faculty and students at Wooster have only limited access to information technology personnel, so they cannot create database-mining software on their own. Biologist Dean Fraga states, "the fact these resources are available free is a great way for smaller institutions to provide students with a taste of this type of research and to access the same data that university researchers have access to."

Using a number of on-line bioinformatics tools, Fraga has created an assignment that is an integral part of his Cell Biology course and provides an opportunity for Biochemistry and Molecular Biology majors to gain the computer expertise that will be needed by future scientists. In this assignment, Fraga has students write a mock

grant proposal[13] for a series of experiments to elucidate the function of a gene in Paramecium based solely on a DNA sequence. In describing the learning outcomes of the assignment, Fraga tells his students, "You will need to develop skill in two areas: searching the scientific literature and bioinformatics. We have constructed this course so that you will develop skills in both."

To accomplish their assignment, students are lead through a series of tasks in which they learn about resources and tools needed to access and analyze sequence data. The students gain an overview of bioinformatics, the concept of genomic database mining, and the development of testable hypotheses on the structure and function of genetic sequences from information gained in other organisms. To accomplish this assignment, students: Download sequence data from Genbank, the DNA sequence database of the National Center for Biotechnology Information (NCBI); Download and use free software that identifies putative open reading frames, the DNA sequences that encode proteins; Use NCBI[14] tools to analyze the predicted open reading frame and discover information known about related genetic sequences.

Fraga's students use these database mining results to "construct a plausible hypothesis as to the role of the identified Paramecium PP1 genes." To do this, students need to use additional web-based NCBI resources. The first is BLAST, a web-based software that uses algorithms that identify Genbank sequences with significant sequence homology. A text-base search and retrieval system, Entrez, is then used to identify other database information (protein sequence, taxonomy, etc.). Another computer-based algorithm, ClustalW, is then used to develop phylogenetic trees based on sequence data. To aid in the analysis of their data, the students are introduced to the mathematical tools of cladistics such as matrix analysis and bootstrapping.

Before writing their mock research proposal, Fraga's students must read the primary research literature related to their sequence. This literature is accessed through Entrez. The students then use computer-generated algorithms to make predictions on the protein folding from their sequence data. Once these assignments are completed, the information is used by the students in the development of their hypothesis concerning the role of the identified Paramecium PP1 genes.

---

[13] www.wooster.edu/biology/dfraga/BIO_305/305_grant_exercise.html
[14] www.ncbi.nlm.nih.gov

The College of Wooster plans to build on their success in bioinformatics undergraduate instruction. "Our philosophy is to embed bioinformatics in four required courses in our biochemistry & molecular biology curriculum rather than develop a new bioinformatics course" reports Fraga. "Our vision is that each course will embed a different aspect of bioinformatics." In the future, Fraga plans to have his students in Cell Physiology actually advance science, by having them study unanalyzed data from the Paramecium genome project.[15] "If I am successful in adding this to my course," observes Fraga, "students will use software for stitching together sequences and identifying coding regions." By addressing the question, "How many genes code for gene X?" undergraduates at the College of Wooster will be helping annotate the Paramecium genome project.

Faculty who are not widely versed in database mining can learn how to use NCBI resources. They provide online instruction manuals and tutorials relating to their resources. Additionally, NCBI will present free workshops on their database tools anywhere in the country, as long as the institution provides 50 attendees.

Contact: Dean Fraga <dfraga@wooster.edu>

Genomics and Computational Biology

Davidson College

Davidson College is a highly selective residential liberal arts college of 1700 students in central North Carolina. Its largest majors are biology, English, and history with about 25% of a graduating class majoring in science or mathematics.

In a two-semester sequence, the Biology and Mathematics Departments at Davidson College provide students with a course on genomics followed by a course on the mathematics needed to understand genomics research. The first course in the sequence is Genomics, Proteomics and Bioinformatics, taught by biologist Malcolm Campbell. The text used for this course is *Discovering Genomics, Proteomics, and Bioinformatics*, coauthored by Campbell and mathematician Laurie

Heyer, who teaches the second course in the series, Computational Biology. According to the course description, students in Campbell's course "will utilize print and online resources to understand how biological information (e.g., DNA sequences, microarrays, proteomics, and clinical studies) is obtained at the genomic level. This information will be integrated into a 'cell web' of molecular interactions."

This first course makes extensive use of a series of "Math Minutes" written by Heyer that are a part of the course textbook. According to Campbell, the Math Minutes are integrated into the textbook "so students can see how the math illuminates the biology. Bioinformatics contains a lot of math and students need to understand the importance of mathematical interpretation of biological data." Campbell also reports that students in this course are required to use their knowledge to answer real-world problems in genomics, proteomics and systems biology (which is a lot of modeling). Through a series of web page assignments, students explore genes and their expressions, along with gene-encoded proteins. Examples of student pages can be viewed on the web.[16]

Computational Biology, which is cross-listed with the mathematics department, is a "survey of bioinformatics techniques used to extract meaning from complex biological data." Heyer has two goals for her students in this course: First, "to understand and apply various algorithms and statistical tests for analyzing DNA, RNA and protein sequences, microarray data, and gene circuits." Second, to "gain practical experience with Perl, a programming language widely used in molecular biology." As part of interdisciplinary teams, students complete two projects that involve building interactive web sites using Perl. The first time the course was offered, students' first project was to describe and demonstrate web sites with Kyte-Doolittle hydropathy plots. Users of these web sites input an amino acid sequence; information on the web page that is produced helps in interpreting the resulting plot. The purpose of the second web project was to introduce hierarchical clustering of gene expression data. For this web site, users input a subset of genes and specify a subset of experimental conditions. Example student projects are on the course web site.[17]

Campbell and Heyer's success is due to several factors. First, both are members of supportive departments. According to Campbell, "neither the College nor our departments had requested that we develop the courses.

[15] www.genoscope.cns.fr/externe/English/Projets/Projet_FN/FN.html

[16] www.bio.davidson.edu/courses/genomics/studentpages.html
[17] www.bio.davidson.edu/courses/compbio/webpage/home.htm

However both departments were keen to support our initiatives." This support came in the form of moral support as well as money to purchase computers and research equipment. In addition, Heyer's department made a commitment to support changes in the curriculum to include more applied mathematics and computational science.

Second, both Campbell and Heyer created their courses using well-developed pedagogical principles. For example, Campbell used his sabbatical to read papers and meet with research-active faculty to enrich the case-based, real-data approach of his course. Heyer created her course around the goals of preparing students to "build new bioinformatics tools" and "develop new approaches to bioinformatics problems." She meets these goals through the use of Perl and a sequence of fundamental bioinformatics algorithms.

Third, their collaborations are an integral part of both courses. "The potential to collaborate with Heyer opened up a new venue for the course," reports Campbell. "Often cell and molecular biology students are told they should study more math, but they are never shown examples of why this is good advice. Heyer was able to identify some interesting biological statements that were founded upon mathematical principles."

Currently, Campbell and Heyer have plans to develop a laboratory genomics course and a genomics concentration that will attract all science majors. "We want to be a funnel rather than a sieve when it comes to attracting students to the intersection of math and biology."

Contacts:
Malcolm Campbell <macampbell@davidson.edu>
Laurie Heyer <laheyer@davidson.edu>

Applying Mathematics in the Sciences

Duquesne University

Located in Pittsburgh, Duquesne is a Catholic university with 5,500 undergraduates and nearly 3,000 graduate students. Business and marketing are the most common major areas but health sciences are second at 17%. The middle 50% of students have SAT scores between 1000 and 1200.

In an attempt to answer the age-old question "why do we have to study all this math?" the mathematics depart-

ment at Duquesne designed a new course for science majors under the broad title Topics in Math. Taught every semester in the mathematics department, it is required of students majoring in biology, chemistry, physics, and physical therapy. Topics in Math is intended to be a capstone mathematics course for science majors that shows them real science applications of mathematics.

The course is taught as a series of three modules, typically four-and-a-half weeks each, covering various topics in mathematics. Each module is geared toward a specific scientific application of mathematics, although the mathematics itself remains the primary emphasis. In the short history of the course, modules have been offered covering such diverse topics as image processing, mathematical modeling, statistical modeling, and knot theory. The topics vary from semester to semester, as do the instructors.

Each of the three modules is designed by a team consisting of one mathematics and one science faculty member. Together, they decide on what to cover, but the mathematician does the instruction. The collaboration goes farther than just course design: typically the science faculty member who has collaborated on design of the module gives one lecture to explain the scientific view on why this mathematical topic is important for science.

Mathematician Eric Rawdon tells of the final assignment where students work in groups of two or three on a project they select from a short list of topics. Each group is assigned a faculty member to work with, and often volunteers from the science departments take part.

"They complete the project and do a short write-up. In their final presentation, they are supposed to apply ideas and techniques from the module to something they have done in a science class. When this isn't possible, the students are allowed to do further research on the subject, the idea being to get them to think beyond the module."

Contact: Eric Rawdon <rawdon@mathcs.duq.edu>

Mathematical Biology

Harvey Mudd College

Harvey Mudd College (HMC) is a small, highly selective, private undergraduate college in southern California with major programs in physics, chemistry, engineering, mathematics, biology, and computer science. The curriculum emphasizes breadth in science and engineering. All HMC students complete a rigorous technical core curriculum that includes courses in all of the above subjects. Students then devote about one-third of their additional course-work efforts to completing major requirements, and another one-third of their course units are focused in the humanities and social sciences. HMC is one of the five colleges in the Claremont Colleges consortium.

HMC offers a new major in mathematical biology that is anchored in two newly designed courses, Mathematical Biology I and II (Math 118 and Math 119),[18] that study mathematical models of biological processes (e.g., population genetics, epidemiology, matrix population models, physiology, and neurobiology). Students in these courses interact with experts in different areas of mathematical biology. Actual topics are determined based on papers written by expert guest lecturers for each semester.

Mathematician Lisette de Pillis and biologist Steve Adolph created these two eight-week courses in 2002 when the new mathematical biology major was established.[19] According to de Pillis, "the main goal of these courses is to expose students to a wide variety of areas in which mathematics and biology are inextricably linked. In order to best do this, we felt students should become acquainted with current research and literature in mathematical biology. This led to the course structure being much like a seminar, in which students read research papers, and experts came to speak on various topics."

In describing the process of developing these courses, de Pillis notes that "Math 118 focuses on continuous dynamic models (PDE and ODE models), while Math 119 focuses on discrete models (including evolutionarily stable systems and discrete dynamic models). With support from a grant that supports the Keck Quantitative Life Sciences Center[20] at Harvey Mudd, several top researchers in mathematical biology were invited to visit HMC and lecture to the class. In addition to arranging for invited guest speakers, we assembled a comprehensive research literature and textbook readings package, and created new sets of lectures."

The prerequisites for the two courses are one semester of Differential Equations and one semester of Linear Algebra along with a course in Introductory Biology. Additional requirements for the mathematical biology major include courses in mathematics, biology and com-

putation. To prepare students for the guest lecturers, both mathematical and biological topics are reviewed or introduced at the beginning of the course. For example, in a recent offering of the first course, students first participated in a workshop on two-species competition models and then learned about partial differential equations. The first guest lecturer, an expert in tumor modeling, then followed. Visits by modeling experts are supplemented in each course with assigned readings, mathematical exercises, and a final project that requires students to review and summarize a pair of articles from research journals in biology or medicine that each present a model of some biological process.

The mathematical biology courses and mathematical biology major are products of the Center for Quantitative Life Sciences at Harvey Mudd. The idea for this Center emerged from discussions among HMC mathematics and biology faculty. In 2000, mathematics department chair Michael Moody wrote a proposal to the Keck Foundation seeking funding to establish a Center. The proposal was funded a year later, and de Pillis and Adolph were named co-directors. The aim of the Center is to introduce students and faculty of HMC and the Claremont Colleges to a "rapidly moving scientific field" through new educational experiences and opportunities for research. In addition to supporting visiting researchers and on-campus research, the Center also promotes course development and provides summer undergraduate research opportunities.

De Pillis and Adolph worked through a number of challenges in implementing their courses, including getting "up to speed in areas outside of our specialties… to help students become ready for the visitors" and not finding a suitable textbook. However, de Pillis reports that, "The most challenging hurdle we have to face … will come in the future: we will have to transition from giving a course which is dominated by outside visitors, to a course which will be taught fully by the two of us."

Contacts:
Lisette de Pillis <depillis@hmc.edu>
Steve Adolph <adolph@hmc.edu>

---

[18] www.math.hmc.edu/~depillis/MATHBIO/index.html

[19] www2.hmc.edu/www_common/biology/academics/biomath.html

[20] www.math.hmc.edu/~depillis/KECK_QLS/index1.html

Mathematical Biology

Hope College

Hope College is a selective residential liberal arts college of 3000 students located in western Michigan. The largest majors are management and psychology, with approximately 50% of graduates having at least one major in the social sciences. The third and fourth largest majors are biology and chemistry, with approximately 20% of graduates having at least one major in the natural science division.

Mathematician Janet Andersen[21] and biologist Leah Chase team-teach Mathematical Biology, a sophomore-level course that is cross-listed in biology and mathematics. The course is based on biology research papers and, according to Andersen, it "fosters interactions between mathematics majors and biology majors that mimic those that occur in interdisciplinary research groups." Objectives outlined in the course description include "communicating across disciplinary boundaries; critically reading research papers; giving oral presentations on technical material to a general audience; learning about areas of research that combine the study of mathematics and biology."

Throughout the course students work in teams to complete both computer labs and wet labs on topics that include population studies, infection dynamics, neuroscience, and animal behavior. Teams have both biology and mathematics student members. "The students work together," says Anderson, "so that the math students help the biology students understand the math and the biology students help the math students understand the biology." The mathematical content covered in the course includes applications in linear algebra and differential equations. Students who elect this course for mathematics credit need to have completed both Linear Algebra and Differential Equations, while the mathematics prerequisite for students seeking biology credit is Calculus I. The prerequisite for biology students is Ecology & Evolution; the mathematics students have no biology prerequisite.

The biology research papers used in the course incorporate both experimental data and mathematical models. Andersen describes a recent unit of the course: "We are currently doing a unit on Hogkin & Huxley's papers based on their research with the squid axon. Students read the Hogkin & Huxley paper, "A quantitative description of membrane current and its application to conduction and excitation in nerve." We present background lectures on cell signaling and analyzing differential equations. We incorporate a wet lab on the isolation and stimulation of the frog sciatic nerve. The students do an oral presentation on the Hodgkin & Huxley paper."

Andersen indicates that the course was developed following a revision of the general education requirements at Hope. "The science requirement for non-majors became a choice of interdisciplinary courses. This began a tradition of faculty in the science division collaborating on course development." Andersen received an NSF grant[22] to support the development of the course which "provided summer stipends for myself, a biologist, and students. It also provided money to hire a part-time instructor, freeing up one of the instructors so that the course could be co-taught."

Commenting on the process of course development, she says, "Two hardest things… have been (1) finding appropriate biology research papers that incorporate mathematical models, and (2) developing the wet labs to accompany the research papers. Most of the faculty summer work was devoted to finding research papers and developing materials to go with them, while most of the student summer work was developing the labs."

"Interdisciplinary conversation is harder and more rewarding than most people think" continued Anderson. "You must learn the vocabulary of the other discipline and develop an understanding of what you don't know in order to effectively communicate."

Mathematical Biology was first offered as a cross-listed course in 2002. It is an elective for both mathematics and biology majors and is offered yearly; typical enrollment is 8 students. In addition to the computer labs and wet labs, students complete a number of oral presentations on their labs and also complete quizzes, homework assignments and discussion questions. All students enrolled in the course must also attend two mathematics and two biology colloquium during the semester, for which they are required to write a brief summary.

Contacts:
Janet Andersen <jandersen@hope.edu>
Leah Chase <chase@hope.edu>

[21] math.hope.edu/andersen/

[22] Course, Curriculum and Lab. Improv. (CCLI) award DUE-0089021.

A New General Education Science Sequence

James Madison University

James Madison University is a largely undergraduate public university in the Shenandoah Valley of Virginia. JMU was founded as a teacher preparation institution but now serves 15,000 students in five colleges.

Faculty at JMU have created a new three-semester package of science courses for students in their Interdisciplinary Liberal Studies (IDLS) program. The courses were created to provide students with a learning experience that models the kind of learning environment that these future teachers will use in their own teaching. The sequence, Understanding our World,[23] consists of six one- or two-credit block courses that provide one option for students in the IDLS program to meet JMU's natural science general education requirements. The last two courses in the sequence, How Life Works—A Human Focus[24] and The Environment in Context, integrate both quantitative and technology applications within the science content.

How Life Works deals with the patterns, energy, information, life's machinery, feedback, community and evolution. Environment in Context uses environmental issues as a unifying concept to introduce ecology, environmental chemistry, and evolution. Topics such as resource utilization and conservation, air and water quality issues, ecological succession, community processes, biological diversity and evolution are used to illustrate concepts and demonstrate the relationship between science and public policy.

According to Cindy Klevickis of JMU's Integrated Science and Technology Department and a member of the development team, "the initiative for this course sequence came from the School of Education as we were developing the IDLS major for future Elementary and Middle School teachers. We wanted students to come away from the class not just knowing science. We wanted them to be able to see science as a possible center for interdisciplinary theme-based learning and ideally we wanted our students to love and appreciate science."

To accomplish this, these IDLS courses are limited to an average class size of 30 so they can provide hands-on,

inquiry-based investigations that are tied to the standards of learning. For example, How Life Works is divided into a series of biology topics, each one of which involves a quantitative or technology component. The topics are:

- Pedigree Workshop
- Ethics and Human Genetic Testing
- Bacteria, Viruses and Infectious Diseases
- Neuromuscular Connections
- Cancer

Technology is a significant component of many of the activities and assignments in Klevickis' courses. "We use Rasmol[25] molecular modeling software to teach the students how to create their own animated computer models of biomolecules."[26] Klevickis adds, "I want my students to be able to use computer applications as a tool to help understand the science content."

Although intended for students in the pre-professional education program, the IDLS program can be selected as a second major for students not in the education program. The Understanding Our World sequence is also taken by some students in the secondary education program and by students who have had negative experiences in the traditional general education science courses.

Contact: Cindy Klevickis < klevicca@jmu.edu>

Calculus and Statistics with Biology Applications

Macalester College

Macalester College is a highly selective liberal arts college located in St. Paul, Minnesota. The college is well known for its large international student population and attracts domestic students from every state in the US. Three of the most popular departments, economics, biology, and psychology, all require introductory statistics as part of their majors. The mathematics and computer science department accounts for about 8% of all declared majors.

In a new two-semester sequence of courses, calculus and statistics are taught at Macalester College using sim-

23 www.isat.jmu.edu/users/klevicca/idls/scicore.htm

24 www.isat.jmu.edu/users/klevicca/idls/GSCI_165.htm

25 www.umass.edu/microbio/rasmol/

26 www.isat.jmu.edu/users/klevicca/vism/vism.htm

ple biological examples. The two courses provide an alternative to the more traditional presentation of these topics, and are the work of mathematicians Daniel Kaplan, Karen Saxe and Tom Halverson. According to Kaplan, "the point of our work has not been to airlift biology examples into a standard curriculum, but to redo the curriculum so that the methods and concepts taught are directly applicable to biology." Biology topics used in these courses range from ecological modeling of single species and of competing/predating species to the action potential of nerve cells.

Beginning with an introduction to the human wake cycle to motivate the need for mathematical models, Calculus with Biology Applications progresses through linear, exponential and logarithmic functions, recursion, and functions of two variables on to derivatives, differential equations, and optimization. To complete a series of computer labs, students make use of the software package "R"[27] made available to them free through the course web site. "The calculus course has been completely redone," reports Kaplan, "de-emphasizing symbolic content and replacing it with geometrical approaches that can be used to analyze the biology applications."

Students also use "R" in Statistics with Biology Applications, which follows the traditional curriculum of introductory statistics courses. Beginning with basic descriptive statistics and graphs, the course continues with probability and probability distributions, regression analysis, statistical inference, analysis of variance, and multiple regression. According to Kaplan, "the most important thing is to emphasize concepts and methods that are directly applicable to the biology as taught by biologists. This means that we don't indulge ourselves in thinking, 'if we teach them how to reason rigorously, the students will somehow figure out how to carry this over to their work in biology.'"

To create these new courses, several steps were taken. First, Kaplan wrote a portion of the biology department's renewal of a grant from the Howard Hughes Medical Institute to provide more advanced mathematics and computer skills for biology majors. In this renewal request, Macalester mathematicians proposed to "revamp the entire introductory mathematics sequence taken by biology majors." Second, the biology department agreed to increase from one to two the number of mathematics department courses required of its majors. To make this change worthwhile to biologists, Kaplan promised that the new required courses would "double

the amount of statistics covered, in addition to giving other mathematical skills that would be useful in studying biology."

Third, Kaplan evaluated the topics in calculus and statistics with an eye towards including only those topics that are important to biology. For more advanced topics, particularly in statistics, he "worked backwards from these to figure out what mathematics the students would need to know." Fourth, with support from the Hughes Institute, Kaplan organized a seminar for biology, mathematics and statistics faculty to go over the topics to be included in the new courses. "The seminar helped to guarantee that the topics would be of interest to the biologists, and created some excitement in the biology department that they would be getting something worthwhile." "At the same time," continued Kaplan, "the seminar helped us justify some seemingly radical steps, such as omitting the topic of symbolic integration almost entirely. The statisticians, I think, were surprised at the sophistication of the statistical techniques that the biologists are interested in."

Originally taught as Special Topics courses, the calculus course will be taught as Math 135 Applied Calculus in the fall of 2004 followed by Math 155 Introduction to Statistical Modeling, pending approval by the College's Educational Policy and Governance Committee.

Contact: Daniel Kaplan <kaplan@macalaster.edu>

Introductory Biology from a Quantitative Approach

Montana State University

Montana State University, a land-grant institution located in Bozeman, offers baccalaureate degrees in 51 fields and doctoral degrees in 17 fields. Enrollment at MSU is over 12,000 students, including undergraduate and graduate students.

Introductory Biology: Cells to Organisms[28] is the first course in a new three-course sequence being offered at Montana State University. The course was developed

[27] www.r-project.org

[28] www.neuron.montana.edu/academics/index.php?fileName=
BIOL213/index.html

under the Hughes Undergraduate Biology (HUB) program[29] at MSU, which was funded by a Howard Hughes Medical Institute grant. According to HUB Program Coordinator Martha Peters, "quantitative approaches are infused in the laboratory exercises, which are original, open-ended labs asking students to explore the experimental process including making hypotheses, designing experiments and analyzing data." Further, the labs are "less content-based and more process-based with a progression of computational approaches used starting from descriptions of measurement and precision to more complex statistical analysis of data."

In the first lab of the course, students study locomotion by measuring the hips and hind limb bones of five cats. Peters says, "In this activity students learn some anatomy and have a discussion of biomechanics, but there is heavy emphasis on measurement and data collection." Summarizing the lab activities she says, "Students are given little direction in how to gather, organize and present their data, but are guided to develop graphs and brainstorm about precision and accuracy of measurement." To complete the lab, students are asked to describe any patterns seen in their data and to verify their results and compare methodologies with other groups in the class.

Offered for the first time in the spring of 2004, Cells to Organisms was taught by Gwen Jacobs from the Cell Biology and Neuroscience Department and Cathy Cripps from Plant Sciences and Plant Pathology. Throughout the course, biology lectures by Jacobs and Cripps are intermixed with lectures on quantitative topics by Neuroscientist Alexander Dimitrov, also from the Cell Biology and Neuroscience Department. For example, following lectures on "Water balance in animals" and "Gas exchange and circulation," Dimitrov lectures on "Uncertainty, measurement and probability" and "Models, relations and graphs."

Materials for Cells to Organisms identify a series of themes that are repeated throughout this first course and its labs and that will be a part of the other two courses in the sequence:

- Scientific processes
- Discovering patterns
- Biological causation
- Balance in dynamic systems
- Levels & scale

---

[29] hughes.montana.edu/curriculum_dev/index.php?fileName=syllabi.html

The other two courses in the sequence, Molecules to Cells and Organisms to Populations, will be taught in the fall of 2004 and the spring of 2005, respectively. The prerequisites for all three courses are the first semester of General Chemistry and either a course in Elementary Statistics or the Survey of Calculus.

According to the HUB web site "the thematic focus of the program is the quantitative analysis of complex biological systems." Course development within HUB is an interdisciplinary endeavor. "No single discipline commands the perspective and depth to surmount the intrinsic challenges in elucidating the structural, organizational, and mechanistic principles of complex biological systems." Faculty from Cell Biology and Neuroscience, Veterinary Molecular Biology, Ecology, Plant Sciences and Plant Pathology and Mathematical Sciences are currently developing new courses under the HUB program.

Contact: Martha Peters <mpeters@cns.montana.edu>

The Many Phases of Mathematical Modeling

Pomona College

One of seven institutions with contiguous campuses in Claremont, California, that comprise The Claremont Colleges, Pomona College is a highly selective institution of about 1,500 students whose median SAT scores are 1450. Biology is one of the top five majors for Pomona students. In addition, 4% major in neuroscience and 4% in mathematics.

Pomona mathematician Ami Radunskaya is always eager to talk about her applied mathematics students' forays into mathematical modeling. "We have developed a six-week module centered around a two-dimensional ODE model of tumor growth with immune competition. Students explore a detailed analysis of this model and show how it can explain some phenomena observed—but not understood—by clinicians, such as tumor regrowth after a long period of remission and asynchronous response to chemotherapy."

This module is part of Mathematical Modeling, a course that guides students as they explore the process of developing, analyzing, simulating, and interpreting scientific data. Although this course is not part of the biology program but rather a requirement of mathematics majors in the "applied math" track, and eighty percent of student projects deal with topics in biology or the life sciences.

Pomona students' work in this course gives evidence that mathematics is present everywhere you find life. Past student projects have included different models of tumor growth (with or without chemotherapy), urban sprawl and the preservation of green space, the behavior of bats in a cave, the synchronization of women's menstrual cycles, the perception of rhythm, and the organization of memory.

But how does a teacher get students to recognize patterns and think about them methodically in a scientific framework? To do so effectively, the instructors concluded that students need a yardstick by which to measure their progress. They have since adopted a detailed approach with a rubric for evaluation provided at each of the following stages: telegraphic reports, project outline, oral presentation, and written presentation.

First, students are asked to search scientific literature for papers on prescribed mathematical models, and to report back to the class in pairs using the "telegraphic report" format. This means they give both a written report and an oral report before the class. Homework assignments are also given, designed to give practice in the analysis of mathematical models.

Second, students present preliminary write-ups for their modeling projects, each of which includes elements such as a project proposal, bibliography, project outline, and poster. Timely feedback is the key at this stage offering possible avenues of improvement early enough for it to be of real help.

Later, students give both an oral and written presentation of their completed project in as professional a manner as possible. "Making the final project presentations a more formal affair encourages the students to take these presentations seriously," Rudanskaya says. "I have seen a distinct improvement in overall output since I started inviting other faculty and students to attend the final presentations."

Contact: Ami Radunskaya <aradunskaya@pomona.edu>

Six Billion People and Counting

Seattle Central Community College

Seattle Central Community College, located in downtown Seattle, serves about 10,000 students per quarter, divided equally into college transfer and professional-technical programs. SCCC was named "College of the Year" in 2001 by Time magazine.

"The main focus of this dual course is to apply mathematics to science, to put mathematics and science together the way they are supposed to be. In this course, mathematics is one of the languages of science, expressing the amount of lead in drinking water in quantitative terms. In this course, mathematics is the tool of science, forecasting how often giant oil spills take place. In this course, mathematics is used to understand environmental issues and problems, and help devise solutions that are based in reality, not appearance."

This is how geologist Joseph Hull and mathematician Greg Langkamp describe the course Six Billion People and Counting[30] they have designed as part of the Coordinated Studies Program (CSP) at Seattle Central Community College. Their course combines introductory environmental science and introductory college-level mathematics in a concentrated learning experience that involves weekly, multi-day projects that are based on real ecological and environmental data and problems. Students receive instruction in the use of Excel for graphing and modeling data, and also take field trips to rivers and forests to gather real data for their examination. At 13 credits, this combined-course experience satisfies both the mathematics/quantitative reasoning and integrated studies requirements of the college's associate of arts degree.

Six Billion People and Counting fits nicely into the philosophy of CSP courses at Seattle Central, which is to give students the opportunity to satisfy several college requirements through the focused integration of topics from different disciplines. According to Hull, the college currently offers both intra- and inter-divisional coordinated studies courses each quarter. Intra-divisional coordinated studies courses within the Science and Mathematics Division are organized under guidelines developed by the Programs Integrating Science and Mathematics (PRISM) committee.

---

[30] seattlecentral.org/faculty/jhull/6bill.html

Hull and Langkamp first started teaching their course as two "linked" but separately taught courses. Starting with lists of 11 mathematics topics and 11 environmental topics, they paired and reorganized the topics to allow a smooth development for both environmental science and mathematics. Multi-day environmentally-based projects are a focus of the course. "We had some project ideas already in mind (e.g., water pollution in a series of lakes modeled with systems of difference equations) that dictated which math and science topics would go together," Hull reports. "Other projects were invented to correspond to the remaining pairs of topics. That was challenging!"

For example, to introduce students early in the semester to studying the relationship between two quantitative variables (as well as to demonstrate the linkages between mathematics and science) 80 shells of the butter clam Saxidomus giganteus were collected for students to measure their lengths and widths. After graphing these data on their TI-83+ calculators, students fit a line first through two selected points and then, using the regression routine, the entire data set. "The purpose of this exercise," says Hull, "is to learn about the real-world application of linear functions, to learn that characteristics of the linear function such as the slope and y-intercept have real physical meaning when applied to real data, and to develop skills in graphically and mathematically representing quantitative information." Students also learn about the random variability of data, outliers, extrapolation, and growth sequences.

Other projects used in this course were generated from classroom materials that are a part of Hull and Langkamp's NSF-funded grant that supported the Quantitative Environmental Learning Project (QELP).[31] Hull and Langkamp are currently preparing a liberal arts mathematics textbook with an environmental theme based on their course.[32]

Even though no quantitative assessment of the impact of this interdisciplinary course on student learning has been made, Hull indicates that the course attracts 40–50 students each time it is offered, and that anecdotal evidence and end-of-class surveys reveal an extremely high satisfaction level. Although the target group for this course is students who are not planning to major in science or mathematics, Hull reports also that they are enrolling an increasing number of science students who are eager to see real applications of mathematics. "With few exceptions, all students felt they learned more math-

ematics when the mathematics was in a socially relevant context and when the mathematics utilized real data and information. A number of liberal arts students from this course have actually gone on to major in science."

Contacts:
Joseph Hull <jhull@sccd.ctc.edu>
Greg Langkamp <glangk@sccd.ctc.edu>

Overcoming Pre-Med Barriers

University of Detroit Mercy

The University of Detroit Mercy is a private, religious, moderately selective master's degree university offering student-centered undergraduate and graduate education in an urban context. The university was established in 1990 by the merger of the Jesuit University of Detroit and Mercy College of Detroit. Today it enrolls about 6000 students in over 100 different academic degrees and programs.

One of the barriers to curriculum reform identified by the *Bio2010* report is the content of high-stakes exams such as the Medical College Admissions Test (MCAT) and the Dental Admissions Test (DAT). Both faculty and students feel that biology courses must teach to these pivotal exams. Time pressure excludes content not directly related to these important evaluations. This creates a dilemma: how to address the curriculum goals of *Bio2010* in a department for which preparation of pre-health professionals is the main educational mission.

The Biology Department of the University of Detroit Mercy (UDM) is just such a department, eager to meet the challenges of *Bio2010* yet constrained by admission tests. The Biology Department offers a broad program to prepare students for graduate or professional school. Approximately 80% of the entering science students are pre-med, pre-dental or pre-physician assistant. The College of Engineering and Science has formal undergraduate programs in each of these health professions.[33]

As a result of the institution's urban location, the Biology Department has diverse demographics. Of the 110 declared biology majors, 75% are female, 17% are African American, 10% are of Middle Eastern descent,

---

[31] seattlecentral.edu/qelp

[32] seattlecentral.edu/qelp/NewBook.html

[33] eng-sci.udmercy.edu/bio/premed.html

5% are Asian, and 4% are Hispanic. The acceptance rate to medical school and dental school is "appreciably above the national average." However, not all students who enter the pre-health profession program enter professional school. Addressing students on its web site, the Biology Department promises that if students decide not to go to medical school after all, faculty will "assist you in revising your plans and getting started in a new direction." As a result of faculty advising, UDM students enter a variety of life sciences related careers, such as biomedical research.

Increasingly, the Biology Department addresses the concerns of *Bio 2010* by introducing quantitative skills into pre-existing courses. "Our students see the value of taking Physiology Laboratory," reports Biology Department chair Greg Grabowski, "but they enter our program with abysmal math skills. To effectively analyze their data, students need to have a firm grasp of statistics, especially parametric tests and standardization of data." To help train students as scientists, the biology department now introduces statistical concepts during the freshmen laboratory experience.

The department introduced statistical thinking in the General Biology Laboratory course as a way to replace cookbook laboratory exercises with inquiry-based instruction. Students learn how to use confidence intervals by measuring the size of brachiopods in fossil casts coming from different rock strata. "Once my students learn how to use this statistical tool [confidence intervals]," reports course instructor Stokes Baker, "they use it in inquiry laboratory exercises." In one exercise, students study the phenomenon of plant cold acclimation, the gaining of freezing resistance once a plant has been exposed to low nonfreezing temperature. To observe this process without specialized equipment, students must compare the growth of acclimated and non-acclimated plant populations. After completing this directed-inquiry laboratory, the students are then challenged to use their newly-acquired statistical skills in an open-ended laboratory investigation. Reporter genes in transgenic plants are used to show that other abiotic stresses, such as drought, induce some of the same genes as cold-stress. The students are assigned an abiotic stress and asked to design and execute an experiment that determines if their assigned stress induced resistance to freezing. "The only way students can answer this question with the resources available to them," observes Baker, "is through the use of statistics."

A formal assessment study of the General Biology Laboratory students showed that incorporating statistics

into an inquiry setting had an impact on learning outcomes. Chi-squared analysis showed that those students given a statistics-based inquiry curriculum, as compared to a course section that used traditional laboratory experiences, were more likely to analyze data involving repeated measurements by calculating a mean ($p = 0.0119$), were more likely to try some method to quantify variability ($p = 9.4 \times 10^{-6}$) and were more likely to correctly deduce that data with excessive overlap cannot be used to make a reasonable conclusion ($p = 0.0122$).

Statistical training is reinforced in a course in Biometrics, which is designed for second year biology and biochemistry students. The rationale in requiring this course for biology majors was that health practitioners are consumers of statistical analysis when they read the scientific literature. Thus UDM's biology graduates should be fluent in these tools even though they are not emphasized on medical entrance exams. Statistical analysis involving standardization of data, parametric analysis, and linear regression are then incorporated into upper level laboratory courses, such as Ecology, Physiology and Animal Behavior.

Modeling using Excel is introduced in upper-level courses. Interactive spreadsheet models are first introduced in the Biometrics course as part of a just-in-time teaching strategy. As Baker observes, "Excel has the advantages of being interactive, readily available and flexible. For example, really complex concepts, like the central limit theorem, can be presented with interactive spreadsheets." Students find modeling an effective tool in their learning. For example, in an anonymous survey with 20 respondents, 80% percent indicated that they agreed or strongly agreed with the statement, "Overall, this demonstration helped me see the concepts in the central limit theorem." 20% were neutral and 0% disagreed.

Students learn how to do modeling by comparing theoretical probability distributions. Baker reports that the skills students learn are "directly transferred to my Ecology course, where students create computer models that simulate predator-prey relationships and logistic growth. With the power and availability of commercial spreadsheets, there is no reason why ecology should be taught purely as a descriptive science."

Students have reported that the addition of mathematics content to biology course has had an impact on their professional lives. One example is Lisa Dosmann, a senior who works part time at Tank Automotive and Armament Command (TACOM), a product development arm of the US Department of Defense. According to Dosmann, she was hired by TACOM, because they were looking for someone with a background other than engineering who also had good research skills. TACOM has asked her to help evaluate the robustness of water quality tests used by soldiers in the field to determine the safety of potential potable water supplies. These tests are being evaluated by receiver operator characteristic (ROC) curves. Referring to her course work in Biometrics, Dosmann said "it was very exiting that we covered what I had to do at work, because ROC curves and regression are integral components of the research at TACOM." "This class [Biometrics] has made me a more valuable employee," she continued. Ms. Dosmann feels that a research-oriented course has been a liberating experience for her. "This is the first class I felt was not pre-med oriented at UDM." Her future plan is to pursue graduate work in biology.

Contact: Stokes Baker <bakerss@udmercy.edu>

Introduction to Mathematical Modeling

University of Mary Washington

The University of Mary Washington is a highly selective residential public liberal arts college of 4000 students in central Virginia. Its largest majors are Business Administration, English, Biology and Psychology; approximately 20% of its graduates major in the natural sciences, mathematics and computer science.

In Introduction to Mathematical Modeling, Mary Washington students learn the process of mathematical modeling by exploring a variety of environmental issues. Recognizing that student interest in the environment was significantly greater than student interest in mathematics, two mathematics faculty—Suzanne Sumner and Debra Hydorn—designed a course that introduces students to the use of mathematical models in understanding environmental issues.

Students first learn to model data with biological and life sciences implications using functions. Sample data sets include mercury concentration in the livers of female dolphins, the number of species of the Galapagos Islands, the number of breeding pairs of Sandwich Terns, and sulphur dioxide pollution potential. Scatter plots of

these and other data sets reveal the need for both linear and non-linear models such as exponential, power, and logarithmic models. The method of least square regression is then presented as a tool for fitting lines to data, along with the log transformation as a means for fitting curves. At each step in the modeling process, students are asked to consider what information a potential model provides about the environment as well as the appropriateness and goodness-of-fit of each model. After finding a "best fit" model for a data set, students are asked to combine what they learned through their model with library and on-line research on the underlying environmental issue to draw conclusions and make recommendations.

Next, students explore the use of difference equations to understand the balance between "restocking" and "harvesting" that is required, for example, in monitoring the population size of species such as Canada Geese, black bears, or zebra mussels in the Great Lakes. Students compare models with different restocking and harvesting requirements and select a best model based on the implications of the environmental situation. The logistic model is then introduced as a modification of the exponential growth model to include an inhibiting factor more appropriate for modeling population growth. The modified logistic model is then explored along with an introduction to chaotic sequences. The final topic typically covered is probability models. Students explore the use of some discrete and continuous probability models, including the exponential and normal distributions, to understand the natural variability within populations.

Course material is supplemented with one or two guest lecturers each semester who provide insights into the usefulness of modeling in understanding environmental problems. Past lecturers include scientist and author John Harte who discussed population models, wildlife photographer and Mary Washington graduate Lynda Richardson who shared her experiences, and geneticist Wyatt Mangum who discussed modeling the impact of habitat fragmentation on a population's gene pool.

In addition to quizzes and exams, students write several papers and complete two group projects. For the papers students examine current research as presented in articles published by NASA's Earth Observatory[34] and then summarize the environmental issues and modeling methods described in the article. Many of the articles presented at this site concern biological or life science research, including "Watching Plants Dance to the

---

[34] earthobservatory.nasa.gov

Rhythms of the Oceans"[35] and "A Delicate Balance: Signs of Change in the Tropics."[36] For the group projects, students complete one data analysis using functions and regression as a modeling tool and one using difference equations. Students use SPSS for the regression project and Excel for the difference equations project.

Sumner and Hydorn have studied the impact of course participation on students' perceptions of their mathematical and analytical abilities. Results thus far suggest an improvement in students' assessment of their abilities and an improved attitude concerning the usefulness of mathematics. Although designed for students who are not planning to major in mathematics or science, the course is enjoying an increased enrollment of mathematics, computer science and science majors, along with students from all disciplines across campus. The course satisfies the general education mathematics requirement at Mary Washington, along with two or more of the college's "across-the-curriculum" requirements (environmental awareness, writing intensive, speaking intensive).

Originally offered as just one section in the Spring of 1998, the mathematics department now offers four sections each semester. Introduction to Mathematical Modeling was developed as part of the Virginia Collaborative for Excellence in the Preparation of Teachers (VCEPT)[37], an NSF-funded program, using interdisciplinary course materials created by Sumner at the 1995 Project Kaleidoscope National Assembly.

Contacts:
  Suzanne Sumner <ssumner@mwc.edu>
  Debra Hydorn < dhydorn@mwc.edu>

Mathematical Consulting

University of Redlands

The University of Redlands is a selective liberal arts university located in southern California. With 2200 students in the residential undergraduate program, and another 2000 enrolled in evening programs in education and business, the University blends a traditional liberal

arts college with offerings in several pre-professional and master's degree programs.

One course at Redlands, Mathematical Consulting,[38] offers students the opportunity to work as part of a team on interdisciplinary projects that link mathematics to problems in environmental and life sciences, economics, and other areas. "On campus, faculty, staff, and administration are able to present problems of academic interest or operational need to the consulting lab; off campus, non-commercial interests and agencies such as schools, social and environmental groups and the police are sources of interesting projects."

The prerequisites for the course are one semester of statistics (selected from courses offered in mathematics, psychology, economics, or business) and permission of the instructor. Additional mathematics and statistics training is provided at the beginning of the course based on the projects for that semester. Much of the first part of the course is also devoted to filling in gaps in the technological preparation of students coming from statistics courses in different disciplines.

Mathematical Consulting was created under Project Intermath[39], an NSF-funded consortium of institutions with the goal of more fully integrating mathematics into the undergraduate curriculum. The course was created by Steve Morics, Rick Cornez and Mike Bloxham, and is currently being taught by Jim Bentley. According to Morics, the course "was created with an eye toward incorporating problems supplied by the life sciences, and several of our students have pursued projects related to life and environmental sciences." Two projects described by Morics include an analysis of bird count data supplied by the local Audubon Society and research with a member of the faculty concerning fish populations in the local mountains. Other projects described on the course web site include "bringing modeling expertise to an EPA-funded effort partly centered on this campus and aimed at saving the Salton Sea" and "assisting in faculty research, such as drawing inferences from aerial imaging in studies of vegetation."

In describing the development of this course, Morics states, "There were two guiding principles. We took as a model a course offered in our environmental sciences program. That course, the Environmental Design Studio, uses teams of students with different strengths to work on actual problems of environmental design. We took

[35] earthobservatory.nasa.gov/Study/SSTNDVI
[36] earthobservatory.nasa.gov/Study/DelicateBalance
[37] vcept.longwood.edu/vcept_course_materials.htm

[38] newton.uor.edu/Departments%26Programs/MathematicsDept/comap/ broch1.html
[39] www.projectintermath.org/

that concept and conceived it as an opportunity to bring students from different majors together working on projects under the guidance of a mathematics professor and an instructor from a project's 'home discipline.' Secondly, we used this as an opportunity to identify what other departments are doing with quantitative information and to try to engage them in the discussion of it. We thought that by inviting them to share their projects with us, they would be more willing to use our expertise in future quantitative endeavors."

The course may be taken for two to four credits; however, according to Steve Morics, "practically every student takes it for four credits, and with a couple of weeks at the front end covering software and other details, research time, and preparing the presentation of the results, most projects take the full 13 weeks of our semester." The course is conducted much like the consulting labs typical of programs in statistics. Although some faculty bringing projects to the course may also recruit students from their disciplines to work on the projects, students who sign up for the course are assigned to a project and will conduct research on the topic of the project and on the methods of analysis needed to complete it. Both students and clients sign a non-disclosure statement and students only report on the result of their analysis without making recommendations to the client.

To prepare for the course, mathematics faculty line up projects from individuals and groups, both on and off campus. "I worked a semester ahead," said Morics. "Our class was new and the campus was full of projects, so it wasn't very hard to drum up business. Indeed, that turned out to be the easiest part of the preparation. The current instructor has far more contacts off-campus than I did, so more projects come in that way, but I don't think he's had to look very far to find them. We offer free consulting to our clients, and they seem fairly eager to take advantage of it." The course is offered on a yearly basis with a typical class size of 10 students.

Contact: Steve Morics <morics@uor.edu>

Making the Microscope a Quantitative Instrument

University of Richmond

The University of Richmond is a comprehensive university in Richmond, Virginia, with about 2800 undergraduates. The Biology Department has 14 full time faculty and graduates 50 majors per year.

The microscope is the symbol of biology because to many people it represents the descriptive nature of this science. Most adults recall that much of their biology education consisted of looking at and drawing objects on microscope slides. However, "anatomy and microscopy can be more than pretty pictures," notes University of Richmond biologist Gary Radice. "Quantitative differences in tissue structure are important and not that difficult to measure."

Thanks to the availability of computer technology, Radice has transformed a traditionally descriptive course, Microanatomy (Bio311), into a highly quantitative course. "My main motivation was that I wanted students to learn to examine images in the microscope critically, and to gather as much information from images as possible," says Radice. "One of the problems novices face is not appreciating the scale of objects they see in the microscope."

In the laboratory component of Microanatomy, students are now asked to compare tissues quantitatively, for example, to compare the capacity and air-exchange areas of frog and mice lungs based on histological sections (microscope slides). One challenge is that images observed from microscope slides are of two-dimensional slices going through complicated three-dimensional tissues; another is that they have no absolute scale. However, it is possible to produce accurate estimates of the volumes of microscopic objects by clever use of statistics and solid geometry. Radice leads his students through a series of activities that teach them how to apply mathematics to histology and then his students use these skill in further inquiry-based investigations.

Stereology is the determination of three-dimensional quantitative attributes of a microscopic specimen by sampling two-dimensional features. Parameters such as size, surface area, volume or volume fraction are estimated using systematic uniform random sampling, a technique where a grid overlay is randomly placed on a microscopic image. Size attributes are then estimated by counting the number of grid points that land within the structure of interest. Once the size of the grid has been calibrated against a standard, equations based on solid geometry are then used to estimate the parameter in question.

The key to stereology is a principle first articulated by the 17th century Jesuit theologian and mathematician Bonaventura Cavalieri: if the equidistant cross sections of two solids with the same altitude are always equal, then the volumes of the two solids are the same. Students gain knowledge about and faith in the accuracy of Cavalieri's principle by determining the volume of yolk and white of a hard boiled egg. Students first determine the volume of their egg by water displacement in a graduated cylinder. Serial sections meeting the requirements of the Cavalieri method are created by using a common kitchen egg-slicer. The egg disks are arranged on a piece of plastic wrap along with a 15 cm length standard. The serial section is then digitized using a common computer scanner.

To measure tissue areas and volumes, grids of regularly spaced crosses are photocopied onto transparent plastic sheets which are then "thrown" onto the images so the grid points fall in a random position. It is then a simple matter to count the number of points that land on white and the number that land on yolk. The area of each square is determined using the scale, and the area of white and yolk on each section is then estimated by multiplying the number of points by the area per square. The volume of the egg is then the sum of the volume of each slice (area times thickness). The calculated volume is then compared to the volume measured by displacement. The method is surprisingly accurate as long as the point grid is placed randomly and the slices are made starting from a random position relative to the end of the egg.

Once students gain skill and confidence in this technique, they then repeat the exercise with commercially available histology serial sections (chicken embryos or frog embryos) or with serial sections the students have prepared as lab projects. The analysis of these microscopic images is accomplished by capturing the slide images with a digital camera attached to a compound microscope. After capturing their images, students transfer them either to CD or over a wireless network to laptop computers. Image analysis is then done using off-the-shelf computer software, either Adobe Photoshop[40] or ImageJ. ImageJ can be downloaded free from the National Institutes of Health (NIH) web site.[41]

Stereology grids can be photocopied onto plastic sheets and applied onto computer monitors or applied onto printed images. Alternately, stereology plug-ins for Photoshop and ImageJ[42] are also available that will scale and draw a randomly-positioned grid of lines, crosses or points directly onto the images. The software method is no more accurate or faster that plastic overlays, but does keep material costs down because large numbers of images do not need to be printed but instead can be analyzed as they are opened on screen. Spreadsheet software is then used to record and statistically analyze the data.

Radice added the quantitative analysis of histological sections as part of his efforts to reform biology education at the University of Richmond. "The general approach I have used in the microanatomy course is to limit the "survey and memorize" aspects of learning tissue structure and focus more on tools and principles. If undergraduates can learn to examine a few organs in depth (e.g., kidney, small intestines, lung) with a variety of tools, they can use those skills later to learn on their own the structure of a liver or some other organ.

As a result of his curriculum reform, Radice has been able to secure National Science Foundation funding to improve his course. The Course Curriculum and Laboratory Improvement (CCLI) program has supported the purchase of imaging equipment and laptop computers for the course. Several web sites[43] and at least one textbook[44] offer good introductions to stereology.

Radice urges others to consider incorporating stereology into their teaching labs. The procedures are simple and easy to learn, the costs of digital microscope cameras have fallen dramatically in recent years, and the necessary computer hardware and software is widely available. "It is a great way to generate data quickly for lessons in basic statistical analysis and experimental design."

Contact: Gary Radice <gradice@richmond.edu>

[40] www.reindeergraphics.com/iptk/

[41] rsb.info.nih.gov/ij/

[42] The most-used plug-ins are "Draw line" and "Point grids."

[43] www.liv.ac.uk/fetoxpath/quantoxpath/stereol.htm
www.stereologysociety.org/
www.reindeergraphics.com/tutorial/index.shtml

[44] Howard, C. V. and M. G. Reed, Unbiased Stereology: Three-Dimensional Measurement in Microscopy, Springer-Verlag, New York, 1998.

Bioinformatics Across the Curriculum

University of Wisconsin—La Crosse

The University of Wisconsin—La Crosse is one of 13 comprehensive universities in the University of Wisconsin System. The University offers associate, bachelor's and master's degree programs, enrolling 8700 undergraduates and 650 master's-level students. The school's most popular majors are business/marketing and social sciences/history; roughly 22% of their students earn a degree in STEM (science, technology, engineering, mathematics) fields.

By creating a series of exercises to be incorporated into existing courses, nine faculty from three different departments established a "Bioinformatics Across the Curriculum" program at the University of Wisconsin-La Crosse. According to Biologist David R. Howard, "the project started with a retreat [whose goal was] to rationally design a program that would introduce students to bioinformatics in a step-wise process throughout their careers." In addition to five biologists, two chemists and two microbiologists are also involved. Howard reports that the faculty involved in the planning phase decided which bioinformatics concepts should be taught in the various courses. A workshop was then given to teach other faculty how to design and incorporate bioinformatics exercises into their courses.

During the summer of 2002, the group created exercises for a total of twelve different courses that span required and elective courses in four disciplines (Biology, Cell and Molecular Biology, Microbiology and Biochemistry). Howard reports that there are numerous activities depending on the level. "In General Biology, students are shown in lecture how sequence alignments are done and used to develop phylogenetic trees. In upper level courses, computational activities include BLAST searches, sequence alignments, protein modeling from x-ray crystal structures, genomics, determining phylogenetic relationships, etc." Of the University's sole Bioinformatics course (Microbiology 440), Howard observes that students "have the opportunity to use virtually every in-silico technique available."

Each bioinformatics exercise created by the group went through a rigorous review process, starting with a presentation to the entire group of participating faculty and an in-depth evaluation by two other faculty. The exercises were then modified based on peer-review and incorporated throughout the academic year. Information about the exercises for five of the courses is available on the Bioinformatics Across the Curriculum project web page[45], which can be accessed through BioWeb[46], a collaborative project among the fourteen different University of Wisconsin System universities and centers to improve undergraduate biology education. The Bioinformatics Across the Curriculum web page is a part of the Molecular genetics resources available under the GenWeb section of BioWeb.[47]

Contact: David Howard <howard.davi@uwlax.edu>

Algorithms and DNA

Wheaton College

Wheaton College is a selective residential liberal arts college of 1600 students in eastern Massachusetts. Its most popular majors are from the social sciences; typically 10% of graduates major in a natural science or mathematics.

Students at Wheaton College have the opportunity to explore genomic research in two separate courses. Algorithms and DNA[48], a required course for computer science majors offered by the Department of Mathematics and Computer Science, is "linked" to a major course in biology. The other, DNA,[49] is a general education, non-science-majors course that is team-taught by faculty in biology and computer science.

Mark LeBlanc, the computer scientist who teaches Algorithms and DNA, says to prospective students: "The sequencing of genomes (e.g., files filled with megabytes of A, C, G, T's) provides a new and rich source of data that is screaming for programmers to help make sense of it all. Said differently, the biologists know that they cannot do the work alone: they need you!"

The value of computer scientists and biologists working together to conduct DNA research is strengthened by

45 bioweb.uwlax.edu/GenWeb/Molecular/ bioinfo%20curric.htm
46 bioweb.uwlax.edu/
47 bioweb.uwlax.edu/GenWeb/Molecular/molecular.htm
48 cs.wheatonma.edu/mleblanc/215/
49 cs.wheatoncollege.edu/mleblanc/dna

linking Algorithms and DNA to one of two courses in the biology department— Genetics or Cell Evolution—that are offered independently but taught at the same time. This "linking" happens as students in the Algorithms and DNA course are paired with a student in one of the biology courses to collaborate in several lab projects throughout the semester and to conduct a final joint research project. This arrangement models a work environment that computer science students will likely encounter in their careers where research and software development is built upon the work of expert scientists.

Toward this end, LeBlanc works closely with biologist Betsey Dexter Dyer who is the instructor of the biology courses to which his course is linked to help define the projects his students will conduct. LeBlanc says this partnership has been invaluable at a small liberal arts college without a formal bioinformatics program. About her collaborations with LeBlanc, Dyer says, "In general we have quite a bit of freedom with constructing our syllabi and we were able to make this connection between our courses without much difficulty." "It helped enormously to be at a small college with one Science Center housing all of the sciences including computer science and mathematics. We (in the Science Center) get to know each other well and share many of the same students. This seems to lead to collaborations."

Algorithms and DNA offers computer science majors an introduction to the mathematical foundations, design, implementation and computational analysis of fundamental algorithms. LeBlanc says, "Algorithms are full of quantitative approaches; it is the essence of the entire course." The emphasis on DNA provides real problems through which students implement procedures such as "heuristic searching, sorting, several graph theory problems, string matching, and the theoretical expression of their orders of growth." This question is most relevant for the biology students as they work with their computer science partner in designing and running a quantitative experiment that searches DNA. One such project required students to design an "Olfactory Gene Finder," i.e., an experiment to search for hydrophobic regions of DNA.

DNA, team-taught by Dyer and LeBlanc, is the second course at Wheaton that gives students opportunity to study the genome. First taught in the fall of 2003 with an enrollment of 18 students, DNA is a general-education elective intended for non-science majors that also satisfies Wheaton's quantitative analysis requirement. The course is designed to show the "amazing blend of biology, chemistry, computing, and mathematics that emerges

when considering DNA." The course description states: "This course explores DNA from four points of view: molecular biology, applied mathematics, organismal and evolutionary biology, and computer science." To conduct DNA analysis, students learn to program in Perl, a string-matching language. Dyer reports that students "design and build (in Perl) 'finders' for genes that possibly code for seven transmembrane proteins." According to Dyer and LeBlanc, student reactions to the course have been "very encouraging." "Anecdotal evidence suggests that many of the non-majors began with misgivings about their ability to program. However, by the end of the semester, many of these same students reported the experience to be 'rewarding' and 'enjoyable.'" Pre- and post-course student evaluations conducted by Dyer and LeBlanc indicate that, "almost the entire class had gained a much greater sophistication in describing topics relevant to DNA."

Student understanding and appreciation for DNA research is further enriched by Wheaton College's curriculum that is centered around the notion of "connections." Wheaton students are required to take at least two connected courses[50] that bridge the traditional divisions. Under the title Genes in Context, Algorithms and DNA and DNA are both connected with the course Ethics, offered by the Philosophy Department. The goal of this connection is "increasing students' awareness and understanding of the ethical issues stemming from the use of our growing knowledge of DNA and the genome."

Dyer and LeBlanc are also the faculty leaders of Wheaton College's Genomics Research project,[51] which provides opportunities for students to contribute to ongoing genomics research. Starting with two student researchers in 1999, this research group now has five student researchers and has produced papers published in *The Journal of Computing Sciences in Colleges and Genome Research.* Dyer and LeBlanc have received grants from NSF for setting up their research lab and for hosting two workshops to help faculty incorporate genomics into the undergraduate curriculum. Recently they received a new NSF award to help disseminate their course and lab materials.

Contacts:
    Betsey Dyer <bdyer@wheatonma.edu>
    Mark LeBlanc <mleblanc@wheatoncollege.edu>

---

[50] www.wheatoncollege.edu/catalog/conx/
[51] genomics.wheatoncollege.edu

Bibliography of Books for Mathematical Biology, Bioinformatics, & Biostatistics

The development and application of quantitative methods in the biological and life sciences have been growing at an enormous rate. This bibliography, long as it is, provides only a very small sample of available resources. Entries were drawn largely from sources cited in the various papers in this volume. For reasons of practicality, the bibliography is limited to books rather than articles; the volume of relevant articles is astronomical. For newcomers, books are more likely than articles to provide introductions and contextual support. We intend no judgment that titles listed here represent the best of their category. Collectively, however, these titles illustrate the enormous penetration of diverse mathematical methods into many important parts of biology and thus offer a type of index to this burgeoning area of interdisciplinary research.

Mathematical Biology (General)
Social Science
Ecological Models
Computing & Biology
Environmental Science & Resource Management
Imaging & Visualization
Evolution
Statistics & Biometrics
Demography & Population Biology
Modeling
Epidemiology
Patterns & Geometry
Medicine
Probability & Combinatorial Mathematics
Physiology
Calculus & Differential Equations
Developmental Biology & Morphogenesis
Stochastic Processes
Molecular and Cellular Biology
Dynamical & Non-Linear Systems
Neuroscience
Oscillators & Reaction-Diffusion Processes
Genetics
Fluids, Waves, Partial Differential Equations
Bioinformatics, Genomics, & Protenomics
Chaos & Fractals
Complexity & Systems Theory
Curriculur & Instructional Resources
Special Biological Models

Mathematical Biology (General)

Bailey, Norman T.J., *The Mathematical Approach To Biology and Medicine,* John Wiley, New York, NY, 1967.

Batschelet, Edward, *Introduction to Mathematics for Life Sciences,* Second Edition, Springer-Verlag, New York, NY, 1976.

Britton, Nicholas F., *Essential Mathematical Biology,* Springer-Verlag, New York, NY, 2003.

Causton, David R., *A Biologist's Mathematics,* Edward Arnold Publishers, London, 1977.

Cullen, Michael R., *Mathematics for the Biosciences,* Prindle Weber & Schmidt, Boston, MA, 1983.

Diekmann, O. and V Capasso, (eds.), *Mathematics Inspired by Biology,* Lecture Notes in Mathematics 1714, Springer-Verlag, New York, NY, 1999.

Eason, G., C.W. Coles, & G. Gettinby, *Mathematics and Statistics for the Bio-Sciences,* Ellis Horwood, West Sussex, UK, 1980.

Geramita, Joan M. and Norman J. Pullman, *An Introduction to the Application of Nonnegative Matrices to Biological Systems,* Queen's University, Kingston, ON, 1984.

Gross, Louis J.and Robert M. Miura, (eds.), *Some Mathematical Questions in Biology: Plant Biology,* Lectures on Mathematics in the Life Sciences, V. 18, American Mathematical Society, Providence, RI, 1986.

Lancaster, H.O., *Quantitative Methods in Biological and Medical Sciences: A Historical Essay,* Springer-Verlag, New York, NY, 1994.

Levin, Simon A., (ed.) *Studies in Mathematical Biology,* 2 Vols., Mathematical Association of America, Washington, DC, 1978.

———, (ed.), *Frontiers in Mathematical Biology,* Lecture Notes in Biomathematics 100. Springer-Verlag, New York, NY, 1994.

Lotka, Alfred J., *Elements of Mathematical Biology,.* Dover, Mineola, NY, 1956.

Murray, James Dickson, *Mathematical Biology,* 2 Vols., Third Edition, Springer-Verlag, New York, NY, 2002.

Newby, J.C., *Mathematics for the Biological Sciences: From Graphs through Calculus to Differential Equations,*Clarendon Press (Oxford), New York, NY, 1980.

Oliveira-Pinto, F. and B.W. Conolly, *Applicable Mathematics of Non-Physical Phenomena,* Halsted Press, New York, NY, 1982.

Rashevsky, Nicholas. *Advances and Applications of Mathematical Biology.* Chicago, IL: University of Chicago Press, 1940.

———, *Mathematical Principles in Biology and Their Applications,* Charles C. Thomas, 1961.

Ricciardi, Luigi and Alwyn Scott, (eds.), *Biomathematics in 1980,* Elsevier North-Holland, New York, NY, 1982.

Rubinow, S.I., *Introduction to Mathematical Biology,* John Wiley, New York, NY, 1975; Dover, Mineola, NY, 2002.

Smith, J. Maynard, *Mathematical Ideas in Biology,* Cambridge University Press, New York, NY, 1968.

Stewart, Ian, *Life's Other Secret: The New Mathematics of the Living World,* John Wiley, New York, NY, 1998.

Yeargers, Edward K., Ronald W. Shonkwiler, & V. Herod, *An Introduction to the Mathematics of Biology with Computer Algebra Models*, Birkhäuser Boston, Boston, MA, 1996.

Ecological Models

Anderson, Roy M. and Robert M. May, *The Dynamics of Human Host-Parasite Systems,* Princeton University Press, Princeton, NJ, 1986.

Bulmer, Michael, *Theoretical Evolutionary Ecology,* Sinauer Associates, Sunderland, MA, 1994.

Cohen, Joel E. *Food Webs and Niche Space.* Princeton, NJ: Princeton University Press, 1978.

Cohen, Joel E., Frédéric Briand, & Charles M. Newman, *Community Food Webs: Data and Theory,.* Biomathematics, V. 20, Springer-Verlag, New York, NY, 1990.

Cohen, Y., (ed.), *Applications of Control Theory in Ecology,* Lecture Notes in Biomathematics 73, Springer-Verlag, New York, NY, 1987.

Diamond, J.M. and T.J. Case, (eds.), *Community Ecology,* Harper & Row, New York, NY, 1986.

Freedman, H.I., *Deterministic Mathematical Models in Population Ecology,* Marcel Dekker, New York, NY, 1980.

Gardner, Robert, (ed.), *Predicting Spatial Effects in Ecological Systems,* Lectures on Mathematics in the Life Sciences, V. 23, American Mathematical Society, Providence, RI, 1993.

Gause, G.F., *The Struggle for Existence,* Williams and Wilkins, Baltimore, MD, 1934; Dover, Mineola, NY, 1971.

Gotelli, Nicholas J., *A Primer of Ecology,* Third Edition, Sinauer Associates, Sunderland, MA, 2001.

Hallam, Thomas G.and Simon A. Levin, (eds.), *Mathematical Ecology, An Introduction,* Springer-Verlag, New York, N, 1986.

Hoff, John and Michael Bevers, *Spatial Optimization in Ecological Applications,* Columbia University Press, New York, NY, 2002.

Jeffries, Clark, *Mathematical Modeling in Ecology: A Workbook for Students,* Birkhäuser Boston, Boston, MA, 1989.

Kot, Mark, *Elements Of Mathematical Ecology,* Cambridge University Press, New York, NY, 2001.

Levin, S.A. and T.G. Hallam, (eds.), *Mathematical Ecology,* Lecture Notes in Biomathematics 54, Springer-Verlag, New York, NY, 1984.

Levin, Simon A., Thomas G. Hallam, & Louis J. Gross, (eds.), *Applied Mathematical Ecology,* Biomathematics, V. 18, Springer-Verlag, New York, NY, 1989.

Logofet, Dmitrii O., *Matrices and Graphs: Stability Problems in Mathematical Ecology,* CRC Press, Boca Raton, FL, 1993.

MacArthur, R.H. and Edward O. Wilson, *The Theory of Island Biogeography,* Princeton University Press, Princeton, NJ, 1967.

May, Robert M. *Stability and Complexity in Model Ecosystems.* Second Edition. Princeton, NJ: Princeton University Press, 1974.

May, Robert M., (ed.), *Theoretical Ecology: Principles and Applications,* Blackwell Science, Malden, MA, 1976.

Nisbet, R.M. and W.S.C. Gurney, *Ecological Dynamics,* Oxford University Press, New York, NY, 1998.

Oster, George F. and Edward O. Wilson, *Caste and Ecology in the Social Insects,* Princeton University Press, Princeton, NJ, 1978.

Pakes, Anthony G. and Ross A. Maller, *Mathematical Ecology of Plant Species Competition,* Studies in Mathematical Biology, Cambridge University Press, New York, NY, 1990.

Patil, Ganapati and Michael L. Rosenzweig, *Contemporary Quantitative Ecology and Related Econometrics,* Statistical Ecology Series, V. 12, International Cooperative Publishers, Fairland, MD, 1979.

Patten, Bernard C.and Sven E. Jo, (eds.), *Complex Ecology: The Part-Whole Relation in Ecosystems,* Prentice Hall, Englewood Cliffs, NJ, 1995.

Pielou, Evelyn C., *Introduction to Mathematical Ecology,* New John Wiley, York, NY, 1970.

Podolsky, Alexander S., *New Phenology: Elements of Mathematical Forecasting in Ecology,* John Wiley, New York, NY, 1984.

Poole, Robert W., *An Introduction To Quantitative Ecology,* McGraw-Hill, New York, NY, 1974.

Rose, Michael R., *Quantitative Ecological Theory: An Introduction to Basic Models,* Johns Hopkins University Press, Baltimore, MD, 1987.

Roughgarden, Jonathan, Robert M. May, & Simon A. Levin, (eds.) *Perspectives in Ecological Theory,* Princeton University Press, Princeton, NJ, 1989.

Roughgarden, Jonathan, *Theory of Population Genetics and Ecology: An Introduction,* Macmillan, New York, NY, 1979.

Schneider, David C., *Primer of Ecological Theory,* Prentice Hall, Upper Saddle River, NJ, 1994.

Scudo, F. and J. Ziegler, *The Golden Age of Theoretical Ecology,* Springer-Verlag, New York, NY, 1978.

Segel, Lee A., *Quantitative Ecology: Spatial and Temporal Scaling,* Academic Press, New York, NY, 1984.

Smith, Hal L. and Paul Waltman, *The Theory of the Chemostat: Dynamics of Microbial Competition,* Studies in Mathematical Biology, Cambridge University Press, New York, NY, 1995.

Smith, J. Maynard, *Models in Ecology,* Cambridge University Press, 1974, New York, NY, 1982.

Tuchinsky, Philip M., *Man in Competition with the Spruce Budworm—An Application of Differential Equations,* Birkhäuser Boston, Boston, MA, 1981.

Turner, Monica G. and Robert H. Gardner, (eds.), *Quantitative Methods in Landscape Ecology: The Analysis and Interpretation of Landscape Heterogeneity,* Springer-Verlag, New York, NY, 1991.

Watt, Kenneth E. F., *Ecology And Resource Management: A Quantitative Approach,* McGraw-Hill, New York, NY, 1968.

Whittaker, R.H., *Communities and Ecosystems,* Second Edition, Macmillan, New York, NY, 1970.

Whittaker, R.H. and Simon A. Levin, *Niche: Theory and Application,* Dowden, Hutchinson and Ross, New York, NY, 1975.

Yodzis, Peter, *Competition for Space and the Structure of Ecological Communities,* Lecture Notes in Biomathematics 25, Springer-Verlag, New York, NY, 1978.

———, *Introduction to Theoretical Ecology,* Addison-Wesley, Reading, MA, 1989.

Environmental Science & Resource Management

Burgman, M.A., S. Ferson, & H. R. Akcakaya, *Risk Assessment in Conservation Biology,* Kluwer, Dordrecht, 1993.

Castillo-Chavez, C., S.A. Levin, & C.A. Shoemaker (eds.), *Mathematical Approaches to Problems in Resource Management and Epidemiology,* Lecture Notes in Biomathematics 81, Springer-Verlag, New York, NY, 1989.

Clark, Colin W., *Bioeconomic Modelling and Fisheries Management,* John Wiley, New York, NY, 1985.

Clark, Colin W., *Mathematical Bioeconomics: The Optimal Management of Renewable Resources,* John Wiley, New York, NY, 1990.

Delic, George and Mary F. Wheeler, (eds.), *Next Generation Environmental Models and Computational Methods,* Society for Industrial and Applied Mathematics, Philadelphia, PA, 1997.

Getz, Wayne M. and Robert G. Haight, *Population Harvesting: Demographic Models of Fish, Forest, and Animal Resources,* Monographs in Population Biology, V. 27, Princeton University Press, Princeton, NJ, 1989.

Haley, K., (ed.), *Applied Operations Research in Fishing,* Plenum Press, New York, NY, 1981.

Lamberson, Roland, (ed.), *Mathematical Models of Renewable Resources,* Humboldt State University, Arcata, CA, 1983.

Mangel, M., (ed.), *Resource Management,* Lecture Notes in Biomathematics 61, Springer-Verlag, New York, NY, 1985.

Manly, Bryan F.J., *Statistics for Environmental Science and Management,* CRC Press, New York, NY, 2000.

McKelvey, Robert W., (ed.), *Environmental and Natural Resource Mathematics,* American Mathematical Society, Providence, RI, 1985.

Rapport, David, *et al.,* (eds.), *Ecosystem Health,* Blackwell Science, Malden, MA, 1998.

Stokes, T.K., J.M. McGlade, & R. Law, (eds.), *Exploitation of Evolving Resources,* Lecture Notes in Biomathematics 99, Springer-Verlag, New York, NY, 1991.

Tilman, D., *Resource Competition and Community Structures,* Princeton University Press, Princeton, NJ, 1982.

Tuljapurkar, Shripad, *Population Dynamics in Variable Environments,* Lecture Notes in Biomathematics 85, Springer-Verlag, New York, NY, 1990.

Vincent, Thomas L., *et al.,* (eds.), *Modeling and Management of Resources under Uncertainty,* Lecture Notes in Biomathematics 72, Springer-Verlag, New York, NY, 1987.

Vincent, Thomas L. and Janislaw M. Skowronski, (eds.), *Renewable Resource Management,* Lecture Notes in Biomathematics 40, Springer-Verlag, New York, NY, 1981.

Waite, T.D. and N.J. Freeman. *Mathematics of Environmental Processes.* Lexington, MA: DC Heath, 1977.

Woodward, F. I. and J.E. Sheehy, *Principles And Measurements In Environmental Biology,* Butterworths, Boston, MA, 1983.

Evolution

Charlesworth, Brian, *Evolution in Age-Structured Populations,* Second Edition, Studies in Mathematical Biology, V.1, Cambridge University Press, New York, NY, 1994.

Charnov, E.L., *The Theory of Sex Allocation,* Princeton University Press, Princeton, NJ, 1982.

Christiansen, F.B. and T.M. Fenchel, (eds.), *Measuring Selection in Natural Populations,* Lecture Notes in Biomathematics 19, Springer-Verlag, New York, NY, 1977.

Feldman, Marcus, (ed.), *Mathematical Evolutionary Theory,* Princeton University Press, Princeton, NJ, 1989.

Galton, Francis, *Natural Inheritance,* Macmillan and Co., London, 1889.

Hofbauer, Josef and Karl Sigmund, *The Theory of Evolution and Dynamical Systems: Mathematical Aspects of Selection,* Cambridge University Press, New York, NY, 1988.

————, *Evolutionary Games and Population Dynamics,* Cambridge University Press, New York, NY, 1998.

Karlin, Samuel and Sabin Lessard, *Theoretical Studies on Sex Ratio Evolution,* Princeton University Press, Princeton, NJ, 1986.

Kirkpatrick, Mark, (ed.), *The Evolution of Haploid-Diploid Life Cycles,* Lectures on Mathematics in the Life Sciences, V. 25, American Mathematical Society, Providence, RI, 1994.

Mangel, Marc, (ed.), *Some Mathematical Questions in Biology: Sex Allocation and Sex Change,* Lectures on Mathematics in the Life Sciences, V. 22, American Mathematical Society, Providence, RI, 1990.

Mani, G.S. *Evolutionary Dynamics of Genetic Diversity,* Lecture Notes in Biomathematics 53, Springer-Verlag, New York, NY, 1984.

Michod, R.E. and B.R. Levin, (eds.), *The Evolution of Sex,* Sinauer Associates, Sunderland, MA, 1988.

Nagylaki, Thomas, *Selection in One- and Two-Locus Systems,* Lecture Notes in Biomathematics 15, Springer-Verlag, New York, NY, 1977.

Ohta, Tomoko, *Evolution and Variation of Multigene Families,* Lecture Notes in Biomathematics 37, Springer-Verlag, New York, NY, 1980.

Smith, John Maynard, *Evolution and the Theory of Games,* Cambridge University Press, New York, NY, 1982.

Demography & Population Biology

Bartlett, M.S., *Stochastic Population Models in Ecology and Epidemiology,* Methuen, New York, NY, 1960.

Bartlett, M.S. and R.W. Hiorns, (eds.), *The Mathematical Theory of the Dynamics of Biological Populations,* Academic Press, New York, NY, 1973.

Brown, Robert L., *Introduction to the Mathematics of Demography,* ACTEX Publications, Winsted, CT, 1991.

Caswell, Hal, *Matrix Population Models Construction, Analysis, and Interpretation,* Second Edition, Sinauer Associates, Sunderland, MA, 2000.

Chapman, Douglas G. and Vincent F. Gallucci, (eds.), *Quantitative Population Dynamics,* Statistical Ecology Series, V. 13, International Cooperative Publishers, Fairland, MD, 1981.

Chiang, Chin Long, *The Life Table And Its Applications,* R.E. Krieger Publishing Company, Malabar, FL, 1984.

Christiansen, F.B. and T.M. Fenchel, *Theories of Populations in Biological Communities,* Springer-Verlag, New York, NY, 1977.

Costantino, Robert F. and Robert A. Desharnais, *Population Dynamics and the Tribolium Model: Genetics and Demography,* Springer-Verlag, New York, NY, 1991.

Cushing, Jim Michael, *An Introduction to Structured Population Dynamics,* Society for Industrial and Applied Mathematics, Philadelphia, PA, 1998.

Frauenthal, James C., *Introduction to Population Modeling,* Birkhäuser Boston, Boston, MA, 1980.

Freedman, H.I. and C. Strobeck, (eds.), *Population Biology,* Lecture Notes in Biomathematics 52, Springer-Verlag, New York, NY, 1983.

Ginzburg, L.R. and E.M. Golenberg, *Lectures in Theoretical Population Biology,* Prentice Hall, Englewood Cliffs, NJ, 1985.

Hassell, Michael P., *The Dynamics of Arthropod Predator-Prey Systems,* Princeton University Press, Princeton, NJ, 1978.

Hastings, Alan, (ed.), *Some Mathematical Questions in Biology: Models In Population Biology,* Lectures on mathematics in the Life Sciences, V. 20, American Mathematical Society, Providence, RI, 1989.

Hastings, Alan, *Population Biology: Concepts and Models,* Springer-Verlag, 1997, New York, NY, 2002.

Hoppensteadt, Frank C., *Mathematical Theories of Populations: Demographics, Genetics, and Epidemics,* Society for Industrial and Applied Mathematics, Philadelphia, PA, 1975.

Hoppensteadt, Frank C., *Mathematical Methods of Population Biology,* Studies in Mathematical Biology, V. 4, New York University Press, 1977; Cambridge University Press, New York, NY, 1982.

Hutchinson, G.E., *An Introduction to Population Ecology,* Yale University Press, New Haven, CT, 1978.

Impagliazzo, J., *Deterministic Aspects of Mathematical Demography,* Springer-Verlag, New York, NY, 1985.

Keyfitz, Nathan and John A. Beekman, *Demography Through Problems,* Springer-Verlag, New York, NY, 1984.

Keyfitz, Nathan, *Introduction to the Mathematics of Population,* Revised Edition, Addison-Wesley, Reading, MA, 1968.

Keyfitz, Nathan, *Applied Mathematical Demography,* John Wiley, New York, NY, 1977.

Kingsland, Sharon E., *Modeling Nature: Episodes in the History of Population Ecology,* University of Chicago Press, Chicago, IL, 1985.

Lewis, E., *Network Models in Population Biology,* Springer-Verlag, New York, NY, 1977.

Ludwig, D., *Stochastic Population Theories,* Springer-Verlag, New York, NY, 1974.

Malthus, Thomas, *An Essay on the Principle of Population,* J. Johnson, St. Paul's Churchyard, London, 1798.

McDonald, L., *et al.,* (eds.), *Estimation and Analysis of Insect Populations,* Lecture Notes in Statistics 55, Springer-Verlag, New York, NY, 1989.

Metz, J.A.J. and O. Diekmann, (eds.), *The Dynamics of Physiologically Structured Populations,* Lecture Notes in Biomathematics 68, Springer-Verlag, New York, NY, 1986.

Mode, C.J., *Stochastic Processes in Demography and Their Computer Implementation,* Biomathematics, V. 14, Springer-Verlag, New York, NY, 1985.

Nisbet, R.M. and W.S.C. Gurney, *Modelling Fluctuating Populations,* John Wiley, New York, NY, 1982; Blackburn Press, 2002.

Pollard, J.H., *Mathematical Models for the Growth of Human Populations,* Cambridge University Press, New York, NY, 1973.

Renshaw, Eric, *Modelling Biological Populations in Space and Time,* Studies in Mathematical Biology, Cambridge University Press, New York, NY, 1991, 1993.

———, *Population and Community Ecology: Principles and Methods,* Cambridge Studies in Mathematical Biology, V. 11, Gordon & Breach, New York, NY, 1991.

Smith, D.P. and Nathan Keyfitz, *Mathematical Demography,* Springer-Verlag, New York, NY, 1978.

Song, Jian and Jingyuan Yu, *Population System Control,* Springer-Verlag, New York, NY, 1988.

Teramoto, E. and M. Yamaguti, (eds.), *Mathematical Topics in Population Biology, Morphogenesis and Neurosciences.* Lecture Notes in Biomathematics 71, Springer-Verlag, New York, NY, 1987.

Therneau, Terry M. and Patricia M. Grambsch, *Modeling Survival Data: Extending the Cox Model,* New York, NY, Springer-Verlag, 2000.

Wilson, Edward O. and William H. Bossert, *A Primer of Population Biology,* Sinauer Associates, Sunderland, MA, 1971.

Epidemiology

Anderson, Roy M. and Robert M. May, (eds.), *Population Biology of Infectious Diseases,* Springer-Verlag, New York, NY, 1982.

Anderson, Roy M., (ed.), *Population Dynamics of Infectious Diseases: Theory and Applications,* Chapman & Hall, New York, NY, 1982.

Asachenkov, Alexander, *et al., Disease Dynamics,* Birkhäuser Boston, Boston, MA, 1994.

Bailey, Norman T.J., *The Mathematical Theory of Infectious Diseases and Its Applications,* Second Edition, Macmillan, New York, NY, 1975.

Brauer, Fred and Carlos Castillo-Chávez, *Mathematical Models in Population Biology and Epidemiology,* Springer-Verlag, New York, NY, 2001.

Busenberg, Stavros and Kenneth Cooke, *Vertically Transmitted Diseases: Models and Dynamics,* Biomathematics, V. 23, Springer-Verlag, New York, NY, 1993.

Castillo-Chavez, C., (ed.), *Mathematical and Statistical Approaches to AIDS Epidemiology,* Lecture Notes in Biomathematics 83, Springer-Verlag, New York, NY, 1989.

Daley, D.J. and J. Gani, *Epidemic Modelling: An introduction,* Studies in Mathematical Biology, Cambridge University Press, New York, NY, 1999, 2001.

Frauenthal, James C., *Mathematical Modeling in Epidemiology,* Springer-Verlag, New York, NY, 1980.

Hethcote, Herbert W. and James A. Yorke, *Gonorrhea, Transmission Dynamics, and Control,* Lecture Notes in Biomathematics 56, Springer-Verlag, New York, NY, 1984.

Hoffmann, G.W. and T. Hraba, (eds.), *Immunology and Epidemiology,* Lecture Notes in Biomathematics 65, Springer-Verlag, New York, NY, 1986.

Kleinbaum, David G., L.L. Kupper & H. Morgenstern, *Epidemiologic Research: Principles and Quantitative Methods,* Lifetime Learning, Belmont, CA, 1982.

Kranz, Jürgen, (ed.), *Epidemics of Plant Diseases: Mathematical Analysis and Modeling,* Second Edition, Springer-Verlag, New York, NY, 1990.

Lauwerier, H.A., *Mathematical Models of Epidemics,* Mathematisch Centrum, Amsterdam, 1981.

Mollison, Denis, (ed.), *Epidemic Models: Their Structure and Relation to Data,* Cambridge University Press, New York, NY, 1995.

Nasell, Ingemar, *Hybrid Models of Tropical Infections,* Lecture Notes in Biomathematics 59, Springer-Verlag, New York, NY, 1985.

Prentice, Ross L. and Alice S. Whittemore, (eds.), *Environmental Epidemiology: Risk Assessment,* Philadelphia, PA, Society for Industrial and Applied Mathematics, 1982.

Medicine

Bailar, John C. and Frederick Mosteller, (eds.), *Medical Uses of Statistics,* Waltham, MA, New England Journal of Medicine Books, 1986.

Bélair, Jacques, *et al.,* (eds.), *Dynamical Disease: Mathematical Analysis of Human Illness,* American Institute of Physics Press, Woodbury, NY, 1995.

Bellman, Richard, *Mathematical Methods in Medicine,* World Scientific, London, 1983.

Bithell, J.F. and R. Coppi, (eds.), *Perspectives in Medical Statistics.* New York, NY: Academic Press, 1981.

Bruni, Carlo, *et al., Systems Theory in Immunology,* Lecture Notes in Biomathematics 32, Springer-Verlag, New York, NY, 1979.

Capasso, V., E. Grosso, & S.L. Paveri-Fontana, (eds.), *Mathematics in Biology and Medicine,* Lecture Notes in Biomathematics 57, Springer-Verlag, New York, NY, 1985.

DeLisi, Charles, *Antigen Antibody Interactions,* Lecture Notes in Biomathematics 8, Springer-Verlag, New York, NY, 1976.

Eisen, Martin M., *Mathematical Models in Cell Biology and Cancer Chemotherapy,* Springer-Verlag, New York, NY, 1979.

Glantz, Stanton A., *Mathematics for Biomedical Applications,* University of California Press, Berkeley, CA, 1979.

Höhne, Karl, (ed.), *Pictorial Information Systems in Medicine,* Springer-Verlag, New York, NY, 1986.

Hoppensteadt, Frank C. and Charles S. Peskin, *Mathematics In Medicine And The Life Sciences,* Springer-Verlag, New York, NY, 1992.

Hoppensteadt, Frank C. and Charles S. Peskin, *Modeling and Simulation in Medicine and the Life Sciences,* Second Edition, Springer-Verlag, New York, NY, 2002.

Ingram, D. and R.F. Bloch, (eds.), *Mathematical Methods in Medicine,* 2 Vols, John Wiley, New York, NY, 1984, 1986.

Iosifescu, M. and P. Tautu, *Stochastic Processes and Applications in Biology and Medicine,* 2 Vols, Springer-Verlag, New York, NY, 1973.

Jacquez, J.A., *Compartmental Analysis in Biology and Medicine,* Second Edition, University of Michigan Press, Ann Arbor, MI, 1985.

Lambrecht, R.M. and A. Rescigno, (eds.), *Tracer Kinetics and Physiologic Modeling,* Lecture Notes in Biomathematics 48, Springer-Verlag, New York, NY, 1983.

Marchuk, Guri I., *Mathematical Modelling of Immune Response in Infectious Diseases,* Kluwer Academic, Dordrecht, 1997.

Miké,Valerie and Kenneth E. Stanley, (eds.), *Statistics In Medical Research: Methods and Issues, with Applications In Cancer Research,* John Wiley, New York, NY, 1982.

Swan, George W., *Applications of Optimal Control Theory in Biomedicine,* Marcel Dekker, New York, NY, 1984.

Umar, Asad, Izet Kapetanovic, & Javed Khan, (eds.), *The Applications Of Bioinformatics In Cancer Detection,* National Institutes of Health, Bethesda, MD, 2002.

Physiology

Alt, W. and G. Hoffmann, (eds.), *Biological Motion,* Lecture Notes in Biomathematics 89, Springer-Verlag, New York, NY, 1991.

Carson, E.R., C. Cobelli, & L. Finkelstein, *The Mathematical Modeling of Metabolic and Endocrine Systems,* John Wiley, New York, NY, 1983.

Childress, Stephen, *Mechanics of Swimming and Flying,* Studies in Mathematical Biology, Cambridge University Press, New York, NY, 1981.

Collins, Richard and Terry J. Van der Werff, *Mathematical Models of the Dynamics of the Human Eye,* Lecture Notes in Biomathematics 34, Springer-Verlag, New York, NY, 1980.

Dallos, P. *et al.,* (eds.), *Mechanics and Biophysics of Hearing,* Lecture Notes in Biomathematics 87, Springer-Verlag, New York, NY, 1990.

Glass, Leon, and Peter Hunter, *et al., Theory of Heart: Biomechanics, Biophysics, and Nonlinear Dynamics of Cardiac Function,* Springer-Verlag, New York, NY, 1991.

Holmes, Mark H.and Lester A. Rubenfeld, (eds.), *Mathematical Modeling of the Hearing Process,* Lecture Notes in Biomathematics 43, New York, NY, Springer-Verlag, 1981.

Hoppensteadt, Frank C., (ed.), *Mathematical Aspects of Physiology,* American Mathematical Society, Providence, RI, 1981.

Keener, James and James Sneyd, *Mathematical Physiology,* Springer-Verlag, New York, NY, 1998.

Lighthill, Michael James, *Mathematical Biofluiddynamics,* Society for Industrial and Applied Mathematics, Philadelphia, PA, 1975.

Mazumdar, J., *et al.,* (ed.), *An Introduction to Mathematical Physiology and Biology,* Second Edition, Studies in Mathematical Biology, Cambridge University Press, New York, NY, 1999.

McMahon, T., *Muscles, Reflexes, and Locomotion,* Princeton University Press, Princeton, NJ, 1984.

Miura, Robert, (ed.), *Some Mathematical Questions in Biology: Muscle Physiology,* Lectures on Mathematics in the Life Sciences, V. 16, American Mathematical Society, Providence, RI, 1986.

Nobel, P., *Biophysical Plant Physiology and Ecology,* W.H. Freeman, New York, NY, 1983.

Panfilov, Alexander V. and Arun V. Holden, (eds.), *Computational Biology of the Heart,* John Wiley, New York, NY, 1997.

Pedley, T.J., ed. *Scale Effects in Animal Locomotion.* New York, NY: Academic Press, 1977.

Strogatz, Steven H., *The Mathematical Structure of the Human Sleep-Wake Cycle,* Lecture Notes in Biomathematics 69, Springer-Verlag, New York, NY, 1986.

Thornley, J.H.M., *Mathematical Models in Plant Physiology,* Academic Press, New York, NY, 1976.

Winfree, Arthur T., *The Timing of Biological Clocks,* Scientific American Library, New York, NY, 1987.

Developmental Biology & Morphogenesis

Bookstein, Fred L., *Measurement of Biological Shape and Shape Change,* Lecture Notes in Biomathematics 24, Springer-Verlag, New York, NY, 1978.

Gilbert, Scott F., *Developmental Biology,* Sinauer Associates, Sunderland, MA, 2003.

Jäger, W., Hermann Rost, & Petre Tautu, *Biological Growth and Spread: Mathematical Theories and Applications,* Springer-Verlag, New York, NY, 1980.

Jean, Roger V., *Phyllotaxis: A Systemic Study in Plant Morphogenesis,* Cambridge University Press, New York, NY, 1994.

Nehaniv, Chrystopher, (ed.), *Mathematical and Computational Biology: Computational Morphogenesis, Hierarchical Complexity, and Digital Evolution,* Lectures on Mathematics in the Life Sciences, V. 26, American Mathematical Society, Providence, RI, 1999.

Prusinkiewicz, Przemyslaw and James Hanan, *Lindenmayer Systems, Fractals, and Plants,* Lecture Notes in Biomathematics 79, Springer-Verlag, New York, NY, 1989.

Rozenberg, G. and A. Salomaa, *The Book of L,* Springer-Verlag, New York, NY, 1986.

Thom, René, *Structural Stability and Morphogenesis: An Outline of a General Theory of Models,* Benjamin Cummings, Redwood City, CA, 1975.

————, *Mathematical Models of Morphogenesis,* Halsted Press, New York, NY, 1983.

Thompson, D'Arcy Wentworth, *On Growth and Form,* Cambridge University Press, Cambridge, UK, 1917, 1942.

Vitanyi, P.M.B., *Lindenmayer Systems: Structure, Languages, and Growth Functions,* Mathematisch Centrum, Amsterdam, 1980.

Williams, R. F., *The Shoot Apex And Leaf Growth:A Study In Quantitative Biology,* Cambridge University Press, New York, NY, 1975.

Molecular and Cellular Biology

Bock, Gregory and Jamie A. Goode, (eds.), *Ion Channels:From Atomic Resolution Physiology To Functional Genomics,* John Wiley, Chichester, UK, 2002.

Burton, T.A., (ed.), *Mathematical Biology: A Conference on Theoretical Aspects of Molecular Science,* Pergamon Press, Elmsford, NY, 1981.

Collado-Vides, Julio, Boris Magasanik, & Temple F. Smith, (eds.), *Integrative Approaches To Molecular Biology,* MIT Press, Cambridge, MA, 1996.

Crippen, G.M. and T.F. Havel, *Distance Geometry and Molecular Conformation,* John Wiley, New York, NY, 1988.

Cronin, Jane, *Mathematics of Cell Electrophysiology,* Marcel Dekker, New York, NY, 1981.

Davis, Rowland H., *The Microbial Models Of Molecular Biology: From Genes To Genomes,* Oxford University Press, New York, NY, 2003.

Fall, Christopher, Eric Marland, John Wagner, and John Tyson, *Computational Cell Biology,* Springer-Verlag, New York, NY, 2002.

Floudas, C.A. and P.M. Pardalos, (eds.), *Optimization in Computational Chemistry and Molecular Biology: Local and Global Approaches,* Kluwer Academic, Dordrecht, 2000.

Goldstein, Byron and Carla Wofsy, (eds.), *Cell Biology,* Lectures on Mathematics in the Life Sciences, V. 24, American Mathematical Society, Providence, RI, 1994.

Gunst, M.C.M. de., *A Random Model for Plant Cell Population Growth,* Mathematisch Centrum, Amsterdam, 1989.

Heinmets, Ferdinand, *Quantitative Cellular Biology; An Approach To The Quantitative Analysis Of Life Processes,* Marcel Dekker, New York, NY, 1970.

Jiang, Tao, Ying Xu, & Michael Q. Zhang, (eds.), *Current Topics in Computational Molecular Biology,* MIT Press, Cambridge, MA, 2002.

Kernevez, Jean-Pierre, *Enzyme Mathematics,* North-Holland, Amsterdam, 1980.

Knolle, Helmut, *Cell Kinetic Modelling and the Chemotherapy of Cancer,* Lecture Notes in Biomathematics 75, Springer-Verlag, New York, NY, 1988.

Macken, Catherine A. and Alan S. Perelson, *Branching Processes Applied to Cell Surface Aggregation Phenomena,* Lecture Notes in Biomathematics 58, Springer-Verlag, New York, NY, 1985.

Macken, Catherine A. and Alan S. Perelson, *Stem Cell Proliferation and Differentiation: A Multitype Branching Process Model,* Lecture Notes in Biomathematics 76, Springer-Verlag, New York, NY, 1988.

McPherson, Alexander, *Introduction to Macromolecular Crystallography,* Wiley-Liss, New York, NY, 2002.

Segel, Lee A., *Modeling Dynamic Phenomena in Molecular and Cellular Biology,* Cambridge University Press, New York, NY, 1984.

Segel, Lee A., (ed.), *Mathematical Models in Molecular and Cellular Biology,* Cambridge University Press, New York, NY, 1980.

Voit, Eberhard, *Computational Analysis of Biochemical Systems,* Cambridge University Press, New York, NY, 2000.

Wood, W.B., *et al., Biochemistry: A Problems Approach,* W.A. Benjamin, Menlo Park, CA, 1974.

Yakovlev, Andrej Yu and Aleksandr V. Zorin, *Computer Simulation in Cell Radiobiology,* Lecture Notes in Biomathematics 74, Springer-Verlag, New York, NY, 1988.

Yockey, Hubert, *Information Theory and Molecular Biology,* Cambridge University Press, New York, NY, 1992.

Neuroscience

Amari, S. and M.A. Arbib, (eds.), *Competition and Cooperation in Neural Nets,* Lecture Notes in Biomathematics 45, Springer-Verlag, New York, NY, 1982.

Arbib, Michael A., *Brains, Machines, and Mathematics,* Second Edition, Springer-Verlag, New York, NY, 1987.

Cronin, Jane, *Mathematical Aspects of Hodgkin-Huxley Neural Theory,* Studies in Mathematical Biology, V. 7, Cambridge University Press, New York, NY, 1987.

Dario, Paolo, (ed.), *Sensors and Sensory Systems for Advanced Robots,* Springer-Verlag, New York, NY, 1988.

Grossberg, Stephen, (ed.), *Neural Networks and Natural Intelligence,* MIT Press, Cambridge, MA, 1988.

Heiden, Uwe an der, *Analysis of Neural Networks,* Lecture Notes in Biomathematics 35, Springer-Verlag, New York, NY, 1980.

Hoppensteadt, Frank C., *An Introduction to the Mathematics of Neurons,* Second Edition, Studies in Mathematical Biology, Cambridge University Press, New York, NY, 1997.

Hoppensteadt, Frank C. and Eugene M. Izhikevich, *Weakly Connected Neural Networks,* Springer-Verlag, New York, NY, 1997.

House, Donald, *Depth Perception in Frogs and Toads: A Study in Neural Computing,* Lecture Notes in Biomathematics 80, Springer-Verlag, New York, NY, 1989.

Kent, Ernest W., *The Brains of Men and Machines,* BYTE/McGraw-Hill, 1981.

MacGregor, Ronald J., *Theoretical Mechanics of Biological Neural Networks,* Academic Press, New York, NY, 1993.

Miura, Robert, (ed.), *Some Mathematical Questions in Biology: Neurobiology,* Lectures on Mathematics in the Life Sciences, V. 15, American Mathematical Society, Providence, RI, 1982.

Peretto, Pierre, *An Introduction to the Modeling of Neural Networks,* Cambridge University Press, New York, NY, 1992.

Rumehlart, David E. and James L. McClelland, *Parallel Distributed Processing: Explorations in the Microstructure of Cognition,* MIT Press, Cambridge, MA, 1986.

Scott, Alwyn C., *Neurophysics,* John Wiley, New York, NY, 1977.

Tahib, Ziad, *Branching Processes and Neutral Evolution,* Lecture Notes in Biomathematics 93, Springer-Verlag, New York, NY, 1992.

Torras, Carme, *Temporal-Pattern Learning in Neural Models,* Lecture Notes in Biomathematics 63, Springer-Verlag, New York, NY, 1985.

Tuckwell, Henry C., *Introduction to Theoretical Neurobiology,* 2 Vols, Studies in Mathematical Biology, Cambridge University Press, New York, NY, 1988.

Tuckwell, Henry, *Stochastic Processes in the Neurosciences,* Society for Industrial and Applied Mathematics, Philadelphia, PA, 1989.

Wu, Jianhong, *Introduction to Neural Dynamics and Signal Transmission Delay,* Walter de Gruyter, Berlin, 2001.

Genetics

Akin, Ethan, *Geometry of Population Genetics,* Lecture Notes in Biomathematics 31, Springer-Verlag, New York, NY, 1979.

Boorman, Scott A. and Paul R. Levitt, *The Genetics of Altruism,* Academic Press, New York, NY, 1980.

Bulmer, Michael G., *The Mathematical Theory of Quantitative Genetics,* Clarendon Press (Oxford), 1980, New York, NY, 1985.

Cannings, Christopher and E.A Thompson, *Genealogical And Genetic Structure,* Cambridge Studies in Mathematical Biology, V. 3, Cambridge University Press, New York, NY, 1981.

Crow, James R. and Moto Kimura, *An Introduction to Population Genetics Theory,* Harper & Row, New York, NY, 1970.

Ewens, Warren J., *Mathematical Population Genetics,* Springer-Verlag, New York, NY, 1979.

Gale, J.S., *Theoretical Population Genetics,* Unwin Hyman, London, 1990.

Gregorius, H.-R., (ed.), *Population Genetics in Forestry,* Lecture Notes in Biomathematics 60, Springer-Verlag, New York, NY, 1985.

Jacquard, Albert, *The Genetic Structure of Populations,* Springer-Verlag, New York, NY, 1974.

Kingman, J.F.C., *Mathematics of Genetic Diversity,* Society for Industrial and Applied Mathematics, Philadelphia, PA, 1980.

Lange, Kenneth, *Mathematical and Statistical Methods for Genetic Analysis,* Springer-Verlag, New York, NY, 1997.

Mather, K. and J.L. Jinks, *Biometrical Genetics,* Chapman & Hall, New York, NY, 1982.

Mirkin, B.G. and S.N. Rodin, *Graphs and Genes,* Biomathematics, V. 11, Springer-Verlag, New York, NY, 1984.

Provine, W., *The Origins of Theoretical Population Genetics,* University of Chicago Press, Chicago, IL, 1971.

Sean B. Carroll, *et al., From DNA to Diversity: Molecular Genetics and the Evolution of Animal Design,* Blackwell Science, Malden, MA, 2001.

Suzuki, David T., Anthony J.F. Griffiths, and Richard C. Lewontin, *An Introduction to Genetic Analysis,* Third Edition, W.H. Freeman, New York, NY, 1976.

Svirezhev, Yuri M. and Vladimir P. Passekov, *Fundamentals of Mathematical Evolutionary Genetics,* Kluwer Academic, Dordrecht, 1990.

Whorz-Busekros, Angelika, *Algebras in Genetics,* Lecture Notes in Biomathematics 36, Springer-Verlag, New York, NY, 1980.

Bioinformatics, Genomics, & Protenomics

Baldi, Pierre and Søren Brunak, *Bioinformatics: The Machine Learning Approach,* MIT Press, Cambridge, MA, 1988, 2001.

Baxevanis, Andreas D., *Current Protocols in Bioinformatics,* Wiley-Interscience, Hoboken, NJ, 2003.

Baxevanis, Andreas D. and B.F. Francis Ouellette, (eds.), *Bioinformatics: A Practical Guide To The Analysis Of Genes And Proteins,* Second Edition, Wiley-Interscience, Hoboken, NJ, 2001.

Bourne, Philip E. and Helge Weissig, (eds.), *Structural Bioinformatics,* Wiley-Liss, Hoboken, NJ, 2003.

Brown, Stuart M., *Bioinformatics: A Biologist's Guide to Biocomputing and the Internet,* Eaton Publishing Company, 2000.

Campbell, A. Malcolm and Laurie J. Heyer, *Discovering Genomics, Proteomics, and Bioinformatics,* Pearson Education, Redwood City, CA, 2002.

Claverie, Jean Michel and Cedric Notredame, *Bioinformatics for Dummies,* Wiley/Dummies, New York, NY, 2003.

Clote, Peter and Rolf Backofen, *Computational Molecular Biology: An Introduction,* John Wiley, New York, NY, 2000.

Durbin, Richard, Sean R. Eddy, Anders Krogh, & Graeme Mitchison, *Biological Sequence Analysis: Probabilistic Models of Proteins and Nucleic Acid,* Cambridge University Press, New York, NY, 1998.

Eidhammer, Ingvar, *et al., Protein Bioinformatics: An Algorithmic Approach to Sequence and Structure Analysis,* John Wiley, New York, NY, 2004.

Ewens, Warren J. and Gregory R. Grant, *Statistical Methods in Bioinformatics,* Springer-Verlag, New York, NY, 2001.

Fasman, D., *Prediction of protein structure and the principles of protein conformation,* Plenum Press, New York, NY, 1989.

Fogel, Gary and David Corne, *Evolutionary Computation in Bioinformatics,* Morgan Kaufman, 2002.

Gibas, Cynthia and Per Jambeck, *Developing Bioinformatics Computer Skills,* O'Reilly, Cambridge, MA, 2001.

Gibson, Greg and Spencer V. Muse, *A Primer Of Genome Science,* Sinauer Associates, Sunderland, MA, 2002.

Gribskov, Michael and John Devereux, *Sequence Analysis Primer,* Oxford University Press, New York, NY, 1994.

Gusfield, Dan, *Algorithms on Strings, Trees, and Sequences: Computer Science and Computational Biology,* Cambridge University Press, New York, NY, 1997.

Ho, Rodney J. Y. and Milo Gibaldi, *Biotechnology And Biopharmaceuticals: Transforming Proteins And Genes Into Drugs,* Wiley-Liss, Hoboken, NJ, 2003.

Innis, Michael A., David H. Gelfand, & John J. Sninsky, (eds.), *PCR Applications: Protocols For Functional Genomics,* Academic Press, San Diego, CA, 1999.

Karwetz, Stephen A. and David D. Womble, *Introduction to Bioinformatics: A Theoretical and Practical Approach,* Humana Press, Totowa, NJ, 2003.

Kohane, Isaac S., Alvin T. Kho, & Atul J. Butte, *Microarrays For An Integrative Genomics,* MIT Press, Cambridge, MA, 2003.

Koski, Timo, *Hidden Markov Models of Bioinformatics,* Kluwer Academic, Dordrecht, 2002.

Krane, Dan E. and Michael L. Raymer, *Fundamental Concepts of Bioinformatics,* Benjamin Cummings, Redwood City, CA, 2003.

Liu, Ben Hui, *Statistical Genomics: Linkage, Mapping, And QTL Analysis,* CRC Press, Boca Raton, FL, 1998.

Misener, Stephen and Stephen A. Krawetz, *Bioinformatics Methods and Protocols,* Humana Press, Totowa, NJ, 2000.

Miura, Robert, (ed.), *Some Mathematical Questions in Biology: DNA Sequence Analysis,* Lectures on Mathematics in the Life Sciences, V. 17, American Mathematical Society, Providence, RI, 1986.

Moody, Glyn, *Digital Code Of Life: How Bioinformatics Is Revolutionizing Science, Medicine, And Business,* Wiley-Interscience, Hoboken, NJ, 2004.

Mount, David W., *Bioinformatics: Sequence And Genome Analysis,* Cold Spring Harbor Laboratory Press, Cold Spring Harbor, NY, 2001.

Percus, Jerome K., *Mathematics of Genome Analysis,* Studies in Mathematical Biology, Cambridge University Press, New York, NY, 2001.

Pevsner, Jonathan, *Bioinformatics and Functional Genomics,* John Wiley, New York, NY, 2003.

Pevzner, Pavel A., *Computational Molecular Biology: An Algorithmic Approach,* MIT Press, Cambridge, MA, 2000.

Rollinson, David and Jennifer Blackwell, *Exploring Parasite Genomes,* Cambridge University Press, New York, NY, 1999.

Sankoff, David and Joseph H. Nadeau, (eds.), *Comparative Genomics: Empirical and Analytical Approaches to Gene Order Dynamics, Map Alignment, and the Evolution of Gene Families,* Computational Biology, v. 1, Kluwer Academic, Dordrecht, 2000.

Stekel, Dov, *Microarray Bioinformatics,* Cambridge University Press, New York, NY, 2003.

Valafar, Faramarz, (ed.), *Techniques In Bioinformatics And Medical Informatics,* New York Academy of Sciences, New York, NY, 2002.

von Baeyer, Hans Christian, *Information: The New Language of Science,* Weidenfeld & Nicolson, 2003.

Waterman, Michael, *Mathematical Methods for DNA Sequences,* CRC Press, Boca Raton, FL, 1989.

Wilkins, M.R., *et al.,* (eds.), *Proteome Research: New Frontiers In Functional Genomics,* New York, NY, Springer-Verlag, 1997.

Complexity & Systems Theory

Aubin, J.-P., D. Saari, & K. Sigmund, (eds.), *Dynamics of Macrosystems,* Lecture Notes in Economics and Mathematical Systems 257, Springer-Verlag, New York, NY, 1985.

Bienenstock, E., *et al. Disordered Systems and Biological Organization,* Springer-Verlag, New York, NY, 1986.

Casti, John L. *Alternate Realities: Mathematical Models of Nature and Man,* John Wiley, New York, NY, 1989.

Casti, John L., *Would-Be Worlds: How Simulation Is Changing the Frontiers of Science.* New York, NY- John Wiley, 1997.

Casti, John L.and Anders Karlqvist, (eds.), *Complexity, Language, and Life: Mathematical Approaches,* Biomathematics, V. 16, New York, NY, Springer-Verlag, 1986.

De Gennes, Pierre-Gilles and Jacques Badoz, *Fragile Objects: Soft Matter, Hard Science and the Thrill of Discovery,* Copernicus (Springer-Verlag), New York, NY, 1996.

DeAngelis, D.L., W.M. Post, & C.C. Travis, *Positive Feedback in Natural Systems,* Biomathematics, V. 15, Springer-Verlag, New York, NY, 1986.

Haken, H., (ed.), *Complex Systems—Operational Approaches in Neurobiology, Physics, and Computers,* Springer-Verlag, New York, NY, 1985.

Heim, Roland. and Günther Palm, (eds.), *Theoretical Approaches to Complex Systems,* Lecture Notes in Biomathematics 21, New York, NY, Springer-Verlag, 1978.

Katz, Michael J., *Templets and the Explanation of Complex Patterns,* New York, NY, Cambridge University Press, 1986.

Mesarovic, M.D. and Y. Takahara, *Abstract Systems Theory,* Lecture Notes in Control and Information Sciences, 116, Springer-Verlag, New York, NY, 1989.

Peak, David and Michael Frame, *Chaos Under Control: The Art and Science of Complexity,* W.H. Freeman, New York, NY, 1994.

Pullman, Bernard, (ed.), *The Emergence of Complexity in Mathematics, Physics, Chemistry and Biology,* Princeton University Press, Princeton, NJ, 1996.

Schuster, P., (ed.), *Stochastic Phenomena and Chaotic Behaviour in Complex Systems,* Springer-Verlag, New York, NY, 1984.

Solé, Richard and Brian Goodwin, *Signs of Life: How Complexity Pervades Biology,* Basic Books, New York, NY, 2001.

Strogatz, Steven H., *Sync: The Emerging Science of Spontaneous Order,* Hyperion, 2003.

Thomas, Renâe, (ed.) *Kinetic Logic: A Boolean Approach to the Analysis of Complex Regulatory Systems,* Lecture Notes in Biomathematics 29, Springer-Verlag, New York, NY, 1979.

Trappl, Robert, (ed.), *Cybernetics: Theory and Applications,* Hemisphere Publishers, Washington, DC, 1983.

Trappl, Robert, George J. Klir, & R. Pichler, (eds.), *Progress in Cybernetics and Systems Research,* Hemisphere Publishers, Washington, DC, 1982.

Special Biological Models

Bukhari, A.I., J. A. Shapiro, & S. L. Adhya, (eds.), *DNA Insertion Elements, Plasmids, And Episomes,* Cold Spring Harbor Laboratory Press, Cold Spring Harbor, NY, 1977.

Dunn, G. and B.S. Everitt, *An Introduction to Mathematical Taxonomy,* Studies in Mathematical Biology, V. 5, Cambridge University Press, New York, NY, 1982.

Levin, Simon A., T.M. Powell, & John H. Steele, (eds.), *Patch dynamics,* Lecture Notes in Biomathematics 96, Springer-Verlag, 1New York, NY, 993.

MacDonald, Norman, *et al.,* (eds.), *Delays in Biological Systems: Linear Stability Theory,* Studies in Mathematical Biology, Cambridge University Press, New York, NY, 1989.

Meek. G.A. and H.Y. Elder, (eds.), *Analytical And Quantitative Methods In Microscopy,* Cambridge University Press, New York, NY, 1977.

Pankhurst, Richard. J., *Practical Taxonomic Computing,* Cambridge University Press, New York, NY, 1991.

Segel, Lee A., (ed.), *Biological Kinetics,* Studies in Mathematical Biology, Cambridge University Press, New York, NY, 1992.

Social Science

Ball, M.A., *Mathematics in the Social and Life Sciences: Theories, Models and Methods,* Halsted Press, New York, NY, 1985.

Bartholomew, David J., *Mathematical Methods in Social Science,* John Wiley, New York, NY, 1981.

Cavalli-Sforza, Luigi Luca, and Marcus W. Feldman, *Cultural Transmission And Evolution:A Quantitative Approach,* Monographs in Population Biology, V. 16, Princeton University Press, Princeton, NJ, 1981.

Dendrinos, Dimitrios S. and Michael Sonis, *Chaos and Socio-Spatial Dynamics,* Springer-Verlag, New York, NY, 1990.

Fararo, Thomas, (ed.), *Mathematical Ideas and Sociological Theory: Current State and Prospects,* Gordon & Breach, New York, NY, 1984.

Gottman, J.D., *et al., The Mathematics of Marriage: Dynamic Nonlinear Models,* MIT Press, Cambridge, MA, 2002.

Iversen, Gudmund R., *Contextual Analysis,* Sage Publishers, Thousand Oaks, CA, 1991.

Raben, Joseph and Gregory Marks, (eds.), *Data Bases in the Humanities and Social Sciences,* North-Holland, Amsterdam, 1980.

Rapoport, Anatol, *Mathematical Models in the Social and Behavioral Sciences,* John Wiley, New York, NY, 1983.

Rashevsky, Nicholas, *Mathematical Theory of Human Relations: An Approach to Mathematical Biology of Social Phenomena,* Second Edition, Principia Press, Bloomington, IN, 1949.

Rashevsky, Nicholas, *Mathematical Biology of Social Behavior,* Revised Edition, University of Chicago Press, Chicago, IL, 1959.

Rothstein, Mark A., (ed.), *Genetic Secrets: Protecting Privacy And Confidentiality In The Genetic Era,* Yale University Press, New Haven, CT, 1997.

Rothstein, Mark A., (ed.), *Pharmacogenomics Social, Ethical, And Clinical Dimensions,* Wiley-Liss, Hoboken, NJ, 2003.

Weidlich, W. and G. Haag, *Concepts and Models of a Quantitative Sociology: The Dynamics of Interacting Populations,* Springer-Verlag, New York, NY, 1983.

Wilson, A.G. *Catastrophe Theory and Bifurcation: Applications to Urban and Regional Systems,* University of California Press, Berkeley, CA, 1981.

Computing and Biology

Bower, James M. and Hamid Bolouri, (eds.), *Computational Modeling of Genetic and Biochemical Networks,* MIT Press, Cambridge, MA, 2001.

Davies, Richard Gareth, *Computer Programming in Quantitative Biology,* Academic Press, New York, NY, 1971.

Filby, Gordon, (ed.), *Spreadsheets in Science and Engineering,* Springer-Verlag, New York, NY, 1998.

Flake, Gary William, *The Computational Beauty of Nature: Computer Explorations of Fractals, Chaos, Complex Systems, and Adaptation,* MIT Press, Cambridge, MA, 1998.

Fortuner, Renaud, (ed.), *Advances in Computer Methods for Systematic Biology: Artificial Intelligence, Databases, Computer Vision,* Johns Hopkins University Press, Baltimore, MD, 1993.

Hameroff, Stuart R., *Ultimate Computing:Biomolecular Consciousness and Nanotechnology,* North-Holland, Amsterdam, 1987.

Jamison, D. Curtis, *Perl Programming for Biologists,* Wiley-Liss, Hoboken, NJ, 2003.

Keen, Robert E. and James D. Spain, *Computer Simulation in Biology: a BASIC Introduction,* Wiley-Liss, New York, NY, 2001.

Kelemenová, Alica and Jozef Kelemen, (eds.), *Trends, Techniques, and Problems in Theoretical Computer Science,* Lecture Notes in Computer Science 281, Springer-Verlag, New York, NY, 1987.

Kunii, Tosiyasu, (ed.), *Frontiers in Computer Graphics,* Springer-Verlag, New York, NY, 1985.

Lee, John David and T.D. Lee, *Statistics And Numerical Methods In BASIC For Biologists,* Van Nostrand Reinhold, New York, NY, 1982.

Mesirov, Jill, (ed.), *Very Large Scale Computation in the 21st Century,* Society for Industrial and Applied Mathematics, Philadelphia, PA, 1991.

Ricciardi, Luigi, (ed.), *Biomathematics and Related Computational Problems,* Kluwer Academic, Dordrecht, 1988.

Simons, Geoff, *The Biology of Computer Life: Survival, Emotion and Free Will,* Birkhäuser Boston, Boston, MA, 1985.

Waterman, Michael S., *Introduction to Computational Biology: Maps, Sequences And Genomes,* Chapman & Hall, New York, NY, 1995.

Wilson, Will. *Simulating Ecological and Evolutionary Systems in C.* New York, NY: Cambridge University Press, 1989, 2000.

Imaging & Visualization

Bushberg, Jerrlod T., *et al.,* (eds.), *The Essential Physics of Medical Imaging,* Lippincott Williams & Wilkins, 2001.

Cheney, Margaret, *et al.,* (eds.), *Tomography, Impedance Imaging, and Integral Geometry,* American Mathematical Society, Providence, RI, 1994.

Cleveland, William S., *Visualizing Data,* Hobart Press, Summit, NJ, 1993.

Deans, Stanley R., *The Radon Transform and Some of Its Applications,* John Wiley, New York, NY, 1983.

Earnshaw, R.A. and N. Wiseman, *An Introductory Guide to Scientific Visualization,* Springer-Verlag, New York, NY, 1992.

Friedhoff, Richard Mark and William Benzon, *The Second Computer Revolution: Visualization,* W.H. Freeman, New York, NY, 1989.

Gerda Kamberova, Shishir Shah, *DNA Array Image Analysis: Nuts & Bolts,* DNA Press, Eagleville, PA, 2002.

Glasbey, Chris A, and G.W. Horgan, *Image Analysis for the Biological Sciences,* John Wiley, New York, NY, 1995.

Gorny, P. and M.J. Tauber, (eds.), *Visualization in Programming,* Lecture Notes in Computer Science 282, Springer-Verlag, New York, NY, 1987.

Hege, Hans-Christian and Konrad Polthier, (eds.), *Visualization and Mathematics: Experiments, Simulations and Environments,* Springer-Verlag, New York, NY, 1997.

Hendee, William R. and E. Russell Ritenour, *Medical Imaging Physics,* Fourth Edition, Wiley-Liss, Hoboken, NJ, 2002.

Herman, Gabor T., *Image Reconstruction from Projections: The Fundamentals of Computerized Tomography,* Academic Press, New York, NY, 1980.

Howard, C. V. and M.G. Reed, *Unbiased Stereology: Three-Dimensional Measurement in Microscopy,* Springer-Verlag, New York, NY, 1998.

National Research Council, *Mathematics and Physics of Emerging Biomedical Imaging,* National Academy Press, Washington, DC, 1996.

Natterer, F., *The Mathematics of Computerized Tomography,* John Wiley, New York, NY, 1986.

Shepp, Lawrence A., *Computed Tomography,* American Mathematical Society, Providence, RI, 1983.

Sklansky, J. and J.-C. Bisconte, (eds.), *Biomedical Images and Computers,* Lecture Notes in Medical Informatics 17, Springer-Verlag, New York, NY, 1982.

Thalmann, Daniel, (ed.), *Scientific Visualization and Graphics Simulation,* John Wiley, New York, NY, 1990.

Wagon, Stan, *The Power of Visualization: Notes from a Mathematica Course,* Front Range Press, 1994.

Statistics & Biometrics

Armitage, Peter and Herbert A. David, (eds.), *Advances in Biometry: 50 Years of the International Biometric Society,* John Wiley, New York, NY, 1996.

Armitage, Peter and Theodore Colton, (eds.), *Encyclopedia Of Biostatistics,* John Wiley, Chichester, UK, 1998.

Bailey, Norman T.J., *Statistical Methods in Biology,* Third Edition, Cambridge University Press, New York, NY, 1995.

Batschelet, Edward, *Circular Statistics in Biology,* Academic Press, New York, NY, 1981.

Berry, Donald A. and Dalene K. Stangl, eds. *Bayesian Biostatistics.* New York, NY: Marcel Dekker, 1996.

Bishop, Owen Neville, *Statistics For Biology: A Practical Guide For The Experimental Biologist,* Longman, Harlow, Essex, UK, 1983.

Blæsild, Preben and Jørgen Granfeldt, *Statistics With Applications In Biology And Geology,* Chapman & Hall/CRC, Boca Raton, FL, 2003.

Campbell, Richard Colin, *Statistics for Biologists,* Third Edition, Cambridge University Press, New York, NY, 1989.

Daniel, Wayne W., *Biostatistics: A Foundation for Analysis in the Health Sciences,* John Wiley, New York, NY, 1991.

Duncan, Robert C., Rebecca G. Knapp, & M. Clinton Miller, *Introductory Biostatistics For The Health Sciences,* John Wiley, New York, NY, 1983.

Dytham, Calvin, *Choosing And Using Statistics: A Biologist's Guide,* Blackwell Science, Malden, MA, 1999.

Federer, Walter T., *Statistical Design and Analysis for Intercropping Experiments: Two Crops,* Springer-Verlag, New York, NY, 1993.

Fernholz, Luisa Turrin, *et al.,* (eds.), *Statistics in Genetics and in the Environmental Sciences,* Birkhäuser Boston, Boston, MA, 2001.

Fisher, Lloyd and Gerald Van Belle, *et al., Biostatistics:A Methodology for the Health Sciences,* Second Edition, Wiley-Interscience, Hoboken, NJ, 2004.

Fleiss, Joseph L., *Statistical Methods for Rates and Proportions,* John Wiley, New York, NY, 1981.

Forthofer, Ronald N. and Eun Sul Lee, *Introduction To Biostatistics: A Guide To Design, Analysis, And Discovery,* Academic Press, San Diego, CA, 1995.

Fry, John C., *Biological Data Analysis: A Practical Approach,* Oxford University Press, New York, NY, 1993.

Gilbert, Neil, *Biometrical Interpretation: Making Sense of Statistics in Biology,* Second Edition, Oxford University Press, New York, NY, 1989.

Gittins, R., *Canonical Analysis: A Review with Applications in Ecology,* Springer-Verlag, New York, NY, 1985.

Glover, Thomas and Kevin Mitchell, *Introduction To Biostatistics,* McGraw-Hill, New York, NY, 2001.

Gotelli, Nicholas J. and Aaron M. Ellison, *A Primer of Ecological Statistics,* Sinauer Associates, Sunderland, MA, 2004.

Harris, Eugene and Albert Adelin, *Survivorship Analysis For Clinical Studies,* Marcel Dekker, New York, NY, 1991.

Heath, David, *An Introduction to Experimental Design and Statistics for Biology,* UCL Press, London, 1995.

Klein, John P. and Melvin L. Moeschberger, *Survival Analysis: Techniques for Censored and Truncated Data,* Springer-Verlag, New York, NY, 1997, 2003.

Kleinbaum, David G., *Logistic Regression: A Self-Learning Text,* Springer-Verlag, New York, NY, 1994.

Kleinbaum, David G., *Survival Analysis: A Self-Learning Text,* Springer-Verlag, New York, NY, 1996.

Kuehl, Robert O., *Statistical Principles of Research Design and Analysis,* Duxbury Press, Pacific Grove, CA, 1994.

Lange, Nicholas, *et al.,* (eds.), *Case Studies in Biometry,* John Wiley, New York, NY, 1994.

Leaverton, Paul E., *A Review Of Biostatistics: A Program For Self-Instruction,* Little, Brown, Boston, MA, 1995.

Lewis, Alvin Edward., *Biostatistics,* Van Nostrand Reinhold, New York, NY, 1984.

Looney, Stephen W., *Biostatistical Methods,* Humana Press, Totowa, NJ, 2002.

Lunneborg, Clifford E., *Modeling Experimental and Observational Data,* Duxbury Press, Pacific Grove, CA, 1994.

Maller, Ross A. and Xian Zhou, *Survival Analysis With Long-Term Survivors,* John Wiley, Chichester, UK, 1996.

Manly, Bryan F.J., *Randomization, Bootstrap And Monte Carlo Methods In Biology,* Second Edition, New York, NY, Chapman & Hall, 1997.

Mead, R., R.N. Curnow, & A.M. Hasted, *Statistical Methods in Agriculture and Experimental Biology,* Second Edition, Chapman & Hall, New York, NY, 1993.

Michalewicz, Zbigniew, (ed.), *Statistical and Scientific Databases,* Ellis Horwood, West Sussex, UK, 1991.

Miller, Rupert G., Jr., *et al.,* (eds.) *Biostatistics Casebook,* John Wiley, New York, NY, 1980.

Milton, Janet Susan, *Statistical Methods in the Biological and Health Sciences,* Second Edition, McGraw-Hill, New York, NY, 1992.

Nelson, Wayne B., *Recurrent Events Data Analysis For Product Repairs, Disease Recurrences, And Other Applications,* Society for Industrial and Applied Mathematics, Philadelphia, PA, 2003.

Neter, John, William Wasserman, & Michael H. Kutner, *Applied Linear Regression Models,* Richard D. Irwin, New York, NY, 1983.

Rosner, Bernard A., *Fundamentals Of Biostatistics,* PWS-Kent, Boston, MA, 1990.

Samuels, Myra L., *Statistics For The Life Sciences,* Dellen Publishing Company, San Francisco, CA, 1989.

Sokal, Robert R. and F. James Rohlf, *Biometry: The Principles and Practice of Statistics in Biological Research,* Second Edition, W.H. Freeman, New York, NY, 1981.

Sokal, Robert R. and F. James Rohlf, *Introduction To Biostatistics,* W.H. Freeman, New York, NY, 1987.

Stroock, D.W., *An Introduction to the Theory of Large Deviations,* Springer-Verlag, New York, NY, 1984.

Tanur, Judith M., *et al.,* (eds.), *Statistics: A Guide to the Study of the Biological and Health Sciences,* Holden-Day, San Francisco, CA, 1977.

Wardlaw, Alastair C., *Practical Statistics for Experimental Biologists,* John Wiley, Chichester, UK, 2000.

Woolson, Robert F., *Statistical Methods for the Analysis of Biomedical Data,* John Wiley, New York, NY, 1987.

Zar, Jerrold H., *Biostatistical Analysis,* Fourth Edition, Prentice Hall, Upper Saddle River, NJ, 1998.

Modeling

Allman, Elizabeth S. and John A. Rhodes, *Mathematical Models in Biology: An Introduction,* Cambridge University Press, New York, NY, 2004.

Avula, Xavier, (ed.), *Mathematical and Computer Modelling in Science and Technology,* Pergamon Press, Elmsford, NY, 1990.

Banks, H.T., *Modeling and Control in the Biomedical Sciences,* Springer-Verlag, New York, NY, 1975.

Barigozzi, Claudio, (ed.), *Vito Volterra Symposium on Mathematical Models in Biology,* Lecture Notes in Biomathematics 39, Springer-Verlag, New York, NY, 1980.

Bossel, Hartmut, *Modeling and Simulation,* AK Peters, Wellesley, MA, 1994.

Brown D. and P. Rothery, *Models in Biology: Mathematics, Statistics and Computing,* John Wiley, New York, NY, 1993.

Caldwell, J. and Y.M. Ram, *Mathematical Modelling: Concepts and Case Studies,* Kluwer Academic, Dordrecht, 1999.

Cherruault, Y., *Mathematical Modelling in Biomedicine: Optimal Control of Biomedical Systems,* D. Reidel Publishers, Norwell, MA, 1985.

Cullen, Michael R., *Linear Models in Biology,* Halsted Press, New York, NY, 1985.

Doucet, Paul and Peter B. Sloep, *Mathematical Modeling in the Life Sciences,* Ellis Horwood, West Sussex, UK, 1992.

Edelstein-Keshet, Leah, *Mathematical Models in Biology,* Random House, New York, NY, 1988.

Finkelstein, Ludwik and Ewart R. Carson, *Mathematical Modelling of Dynamic Biological Systems,* Research Studies Press, Forest Grove, OR, 1979.

France, J. and J. H. M. Thornley, *Mathematical Models in Agriculture,* Butterworths, London, 1984.

Haefner, James W., *Modeling Biological Systems: Principles and Applications,* Kluwer Academic, Dordrecht, 1996.

Lunneborg, Clifford E., *Modeling Experimental and Observational Data,* Duxbury Press, Belmont, CA, 1994.

Marcus-Roberts, Helen and Maynard Thompson, (eds.), *Life Science Models: Modules in Applied Mathematics,* Springer-Verlag, New York, NY, 1983.

Mooney, Douglas D. and Randall J. Swift, *A Course in Mathematical Modeling,* Mathematical Association of America, Washington, DC, 1999.

Othmer, H. G., *et al.,* (eds.), *Case Studies in Mathematical Modeling: Ecology, Physiology and Cell Biology,* Prentice Hall, Upper Saddle River, NJ, 1997.

Roberts, Fred S., *Discrete Mathematical Models with Applications to Social, Biological, and Environmental Problems,* Prentice Hall, Englewood Cliffs, NJ, 1976.

Rodin, Ervin Y.and Xavier J.R. Avula, (eds.), *Mathematical Modelling in Science and Technology,* Pergamon Press, Elmsford, NY, 1988.

Saaty, Thomas L. and Joyce M. Alexander, *Thinking With Models: Mathematical Models in the Physical, Biological, and Social Sciences,* Pergamon Press, Elmsford, NY, 1981.

Solomon, D.L. and C. Walter, (eds.), *Mathematical Models in Biological Discovery,* Lecture Notes in Biomathematics 13, Springer-Verlag, New York, NY, 1977.

Starfield, Anthony and A. L. Bleloch, *Building Models for Conservation and Wildlife Management,* Second Edition, Interacton Books, New York, NY, 1991.

Thornley, John H. M. and Ian R. Johnson, *Plant and Crop Modeling,* Clarendon Press (Oxford), New York, NY, 1990.

Patterns & Geometry

Adam, John A., *Mathematics in Nature: Modeling Patterns in the Natural World,* Princeton University Press, Princeton, NJ, 2003.

Albrecht, Duane G., (ed.), *Recognition of Pattern and Form,* Lecture Notes in Biomathematics 44, Springer-Verlag, New York, NY, 1982.

Bookstein, Fred L., *Morphometric Tools for Landmark Data: Geometry and Biology,* Cambridge University Press, New York, NY, 1991.

Carpenter, Gail, (ed.), *Some Mathematical Questions in Biology: Circadian Rhythms,* Lectures on Mathematics in the Life Sciences, V. 19, American Mathematical Society, Providence, RI, 1987.

Cook, Theodore Andrea, *The Curves of Life,* Dover, Mineola, NY, 1979.

Cosnard, M., J. Demongeot, & A. Le Breton, (eds.), *Rhythms in Biology and Other Fields of Application: Deterministic and Stochastic Approaches,* Lecture Notes in Biomathematics 49, Springer-Verlag, New York, NY, 1983.

Findley, A.M., S.P. McGlynn, & G.L. Findley, *The Geometry of Genetics,* John Wiley, New York, NY, 1989.

Glass, Leon and M.C. Mackey, *From Clocks to Chaos: The Rhythms of Life,* Princeton University Press, Princeton, NJ, 1988.

Jäager, W. and J.D. Murray, (eds.), *Modelling of Patterns in Space and Time,* Lecture Notes in Biomathematics 55, Springer-Verlag, New York, NY, 1984.

Kaandorp, Jap and Janet Kübler, *The Algorithmic Beauty of Seaweeds, Sponges and Corals,* Springer-Verlag, New York, NY, 2001.

MacDonald, Norman, *Time Lags in Biological Models,* Lecture Notes in Biomathematics 27, Springer-Verlag, New York, NY, 1978.

Maini, Philip K. and Hans G. Othmer, (eds.), *Mathematical Models for Biological Pattern Formation,* Springer-Verlag, New York, NY, 2000.

Todd, Philip H., *Intrinsic Geometry of Biological Surface Growth,* Lecture Notes in Biomathematics 67, Springer-Verlag, New York, NY, 1986.

Williams, Carrington Bonsor, *Patterns in the Balance of Nature and Related Problems in Quantitative Ecology,* Academic Press, New York, NY, 1964.

Winfree, Arthur T., *The Geometry of Biological Time,* Second Edition, Springer-Verlag, New York, NY, 1980, 1990, 2001.

Probability & Combinatorial Mathematics

Berg, Howard C., *Random Walks in Biology,* Princeton University Press, Princeton, NJ, 1983, 1993.

Dale, Andrew I., *A History of Inverse Probability: From Thomas Bayes to Karl Pearson,* Springer-Verlag, New York, NY, 1991.

Denny, Mark and Stephen Gaines, *Chance in Biology: Using Probability to Explore Nature,* Princeton University Press, Princeton, NJ, 2000, 2002.

Durret, Richard, *Probability Models for DNA Sequence Evolution,* Springer-Verlag, New York, NY, 2002.

Edwards, A.W.F., *Pascal's Arithmetical Triangle,* Oxford University Press, New York, NY, 1987.

Gigerenzer, Gerd, *et al., The Empire of Chance: How Probability Changed Science and Everyday Life,* Cambridge University Press, New York, NY, 1989.

Jagers, P., *Branching Processes with Biological Applications,* John Wiley, New York, NY, 1975.

Kemeny J. G, Snell J. L., and Thompson G. L., *Introduction to Finite Mathematics,* Prentice Hall, Upper Saddle River, NJ, 1956.

Kimmel, Marek and David E. Axelrod, *Branching Processes In Biology,* Springer-Verlag, New York, NY, 2001.

Krüger, Lorenz, *et al.,* (eds.), *The Probabilistic Revolution,* MIT Press, Cambridge, MA, 1987.

MacDonald, Norman, *Trees and Networks in Biological Models,* John Wiley, New York, NY, 1983.

Miles, R. E. and J. Serra, (eds.), *Geometrical Probabilities and Biological Structures: Buffon's 200th Anniversary,* Lecture Notes in Biomathematics 23, Springer-Verlag, New York, NY, 1978.

Percus, J.K., *Combinatorial Methods in Developmental Biology,* Courant Institute of Mathematical Sciences, New York, NY, 1977.

Calculus & Differential Equations

Abell, Martha L. and James P. Braselton, *Modern Differential Equations: Theory, Applications, Technology,* Saunders College, Philadelphia, PA, 1996.

Adler, Frederick R., *Modeling the Dynamics of Life: Calculus and Probability for Life Scientists,* Brooks/Cole, Pacific Grove, CA, 1998.

Borrelli, Robert L. and Courtney S. Coleman, *Differential Equations: A Modeling Perspective,* John Wiley, New York, NY, 1998.

Braun, Martin, Courtney S. Coleman, & A. Drew, (eds.), *Differential Equation Models,* Springer-Verlag, New York, NY, 1983.

Burghes, D.N. and M.S. Borrie, *Modelling with Differential Equations,* Halsted Press, New York, NY, 1981.

Burton, T.A., (ed.), *Modeling and Differential Equations in Biology,* Marcel Dekker, New York, NY, 1980.

Busenberg, S. M. Martelli, (ed.), *Differential Equations Models in Biology, Epidemiology, and Ecology,* Lecture Notes in Biomathematics 92, Springer-Verlag, New York, NY, 1991.

Busenberg, Stavros and Kenneth Cooke, (eds.), *Differential Equations and Applications in Ecology, Epidemics, and Population Problems,* Academic Press, New York, NY, 1981.

Clow, Duane J. and N. Scott Urquhart, *Mathematics in Biology: Calculus and Related Topics,* Ardsley House, Lanham, MD, 1984.

Cornish-Bowden, Athel, *Basic Mathematics for Biochemists,* Chapman & Hall, New York, NY, 1981.

Cronin, Jane, *Differential Equations: Introduction and Qualitative Theory,* Second Edition, Marcel Dekker, New York, NY, 1994.

Ellis, Wade, *et al., Calculus: Mathematics and Modeling,* Addison Wesley/Pearson, Reading, MA, 1999.

Jones, D.S. and B.D. Sleeman, *Differential Equations and Mathematical Biology,* Allen & Unwin, Sydney, 1983; Chapman & Hall/CRC, Boca Raton, FL, 2003.

Neuhauser, Claudia, *Calculus for Biology and Medicine,* Second Edition, Prentice Hall, Upper Saddle River, NJ, 2000.

Taubes, Clifford Henry, *Modeling Differential Equations in Biology,* Prentice Hall, Upper Saddle River, NJ, 2001.

Stochastic Processes

Alstad, Don, *An Introduction to Stochastic Processes with Applications to Biology,* Pearson Prentice Hall, Upper Saddle River, NJ, 2001.

Gabriel, J.-P., C. Lefaevre, P. Picard, (eds.), *Stochastic Processes in Epidemic Theory,* Lecture Notes in Biomathematics 86, Springer-Verlag, New York, NY, 1989.

Gilks, W.R., S. Richardson, & D.J. Spiegelhalter, (eds.), *Markov Chain Monte Carlo In Practice,* Chapman & Hall, New York, NY, 1996.

Goel, Narenda S. and Nira Richter-Dyn. *Stochastic Models in Biology.* New York, NY: Academic Press, 1974.

Kimura, Motoo, G. Kallianpur, & Takeyuki Hida, (eds.), *Stochastic Methods in Biology,* Lecture Notes in Biomathematics 70, Springer-Verlag, New York, NY, 1987.

Tautu, P., (ed.), *Stochastic Spatial Processes,* Lecture Notes in Mathematics 1212, Springer-Verlag, New York, NY, 1986.

Yoshimura, Jin and Colin W. Clark, (eds.), *Adaptation in Stochastic Environments,* Lecture Notes in Biomathematics 98, Springer-Verlag, New York, NY, 1993.

Dynamical and Non-Linear Systems

Adler, Fred, *Dynamics of Life,* Brooks/Cole, Pacific Grove, CA, 1998.

Arrowsmith, D.K. and C.M. Place, *An Introduction to Dynamical Systems,* Cambridge University Press, New York, NY, 1990.

Aston, Philip, ed. *Nonlinear Mathematics and Its Applications.* New York, NY: Cambridge University Press, 1996.

Basseville, M. and A. Benveniste, (eds.), *Detection of Abrupt Changes in Signals and Dynamical Systems,* Lecture Notes in Control and Information Sciences 77, Springer-Verlag, New York, NY, 1986.

Brokate, Martin and Jürgen Sprekels, *Hysteresis and Phase Transitions,* Springer-Verlag, New York, NY, 1996.

DeAngelis, D. L and L. J. Gross, (eds.), *Dynamics of Nutrient Cycling and Food Webs,* Chapman & Hall, New York, NY, 1992.

Demongeot, J., E. Goles, & M. Tchuente, (eds.), *Dynamical Systems and Cellular Automata,* Academic Press, New York, NY, 1985.

Enns, Richard H., *et al.,* (eds.), *Nonlinear Phenomena in Physics and Biology,* Plenum Press, New York, NY, 1981.

Hastings, Alan, *Dynamic Modeling,* Springer-Verlag, New York, NY, 1997.

Kaplan, Daniel and Leon Glass. *Understanding Nonlinear Dynamics.* New York, NY: Springer-Verlag, 1995.

Mangel, Marc and Colin W. Clark, *Dynamic Modeling in Behavioral Ecology,* Princeton University Press, Princeton, NJ, 1988.

Murray, J.D., *Lectures on Nonlinear-Differential-Equation Models in Biology,* Clarendon Press (Oxford), New York, NY, 1977.

Othmer, Hans, (ed.), *Some Mathematical Questions in Biology: The Dynamics of Excitable Media,* Lectures on Mathematics in the Life Sciences, V. 21, American Mathematical Society, Providence, RI, 1989.

Ruelle, David, *Elements of Differentiable Dynamics and Bifurcation Theory,* Academic Press, New York, NY, 1989.

Smítalová, Kristína and Stefan Sujan, *A Mathematical Treatment of Dynamical Models in Biological Science,* Ellis Horwood, New York, NY, 1991.

West, Bruce J., *An Essay on the Importance of Being Nonlinear.* Lecture Notes in Biomathematics 62, Springer-Verlag, New York, NY, 1985.

Oscillators & Reaction-Diffusion Processes

Banks, Robert B., *Growth and Diffusion Phenomena: Mathematical Frameworks and Applications,* Springer-Verlag, New York, NY, 1994.

Fife, Paul C., *Mathematical Aspects of Reacting and Diffusing Systems,* Lecture Notes in Biomathematics 28, Springer-Verlag, New York, NY, 1979.

Graef, John R. and Jack K. Hale, (eds.), *Oscillation and Dynamics in Delay Equations,* American Mathematical Society, Providence, RI, 1992.

Grasman, Johan, *Asymptotic Methods for Relaxation Oscillations and Applications,* Springer-Verlag, New York, NY, 1987.

Hodgson, J.P.E., (ed.), *Oscillations in Mathematical Biology,* Lecture Notes in Biomathematics 51, Springer-Verlag, New York, NY, 1983.

Hoppensteadt, Frank C., (ed.), *Nonlinear Oscillations in Biology,* American Mathematical Society, Providence, RI, 1979.

Neimark, Yu. I. and P.S. Landa, *Stochastic and Chaotic Oscillations,* Kluwer Academic, Dordrecht, 1992.

Othmer, H.G., (ed.), *Nonlinear Oscillations in Biology and Chemistry,* Lecture Notes in Biomathematics 66, Springer-Verlag, New York, NY, 1986.

Pavlidis, Theo, *Biological Oscillators: Their Mathematical Analysis,* Academic Press, New York, NY, 1973.

Rothe, Franz, *Global Solutions of Reaction-Diffusion Systems,* Lecture Notes in Mathematics 1072, Springer-Verlag, New York, NY, 1984.

Sachdev, P.L., *Nonlinear Diffusive Waves,* Cambridge University Press, New York, NY, 1987.

Tyson, John J., *Belousov-Zhabotinskii reaction,* Lecture Notes in Biomathematics 10, Springer-Verlag, New York, NY, 1976.

Fluids, Waves, & Partial Differential Equations

Aldroubi, Akram and Michael Unser, (eds.), *Wavelets in Medicine and Biology,* CRC Press, Boca Raton, FL, 1996.

Barenblatt, Grigory Isaakovich, *Scaling Phenomena in Fluid Mechanics,* Cambridge University Press, New York, NY, 1995.

Boccardo, L. and A. Tesei, (eds.), *Nonlinear Parabolic Equations: Qualitative Properties of Solutions,* Longman Scientific, London, 1987.

Cheer, A.Y. and C.P. van Dam, (eds.), *Fluid Dynamics In Biology,* American Mathematical Society, Providence, RI, 1993.

Davis, Julian L., *Wave Propagation in Solids and Fluids,* Springer-Verlag, New York, NY, 1988.

Davydov, A.S., *Solitons in Molecular Systems,* Second Edition, Kluwer Academic, Dordrecht, 1991.

Fitzgibbon, III, W.E., (ed.), *Partial Differential Equations and Dynamical Systems,* Pitman, 1984.

Grusa, K-U., *Mathematical Analysis of Nonlinear Dynamic Processes,* Longman Scientific, London, 1988.

Hirose, Akira and Karl E. Lonngren, *Introduction to Wave Phenomena,* John Wiley, New York, NY, 1985.

Infeld, E. and G. Rowlands, *Nonlinear Waves, Solitons and Chaos,* Cambridge University Press, New York, NY, 1990.

Pedley, T.J., *The Fluid Mechanics of Large Blood Vessels,* Cambridge University Press, New York, NY, 1980.

Ricciardi, L.M., *Diffusion Processes and Related Topics in Biology,* Springer-Verlag, New York, NY, 1977.

Vogel, S., *Life in Moving Fluid: The Physical Biology of Flow,* Princeton University Press, Princeton, NJ, 1983.

Chaos & Fractals

Berry, M.V., I.C. Percival, & N.O. Weiss, (eds.), *Dynamical Chaos,* Princeton University Press, Princeton, NJ, 1987.

Fischer, P. and William R. Smith, (eds.), *Chaos, Fractals, and Dynamics,* Marcel Dekker, New York, NY, 1985.

Fleischmann, M., D.J. Tildesley, & R.C. Ball, (eds.), *Fractals in the Natural Sciences,* Princeton University Press, Princeton, NJ, 1990.

Grebogi, Celso and James A. Yorke, (eds.), *The Impact of Chaos on Science and Society,* United Nations University Press, New York, NY, 1997.

Hastings, Harold M. and George Sugihara, *Fractals: A User's Guide for the Natural Sciences,* Oxford University Press, New York, NY, 1993.

Holden, Arun, (ed.), *Chaos,* Princeton University Press, Princeton, NJ, 1986.

Mandelbrot, Benoit, *Les objets fractals: forme, hazard et dimension,* Flammarion, Paris, 1975.

Marek, Milos and Igor Schreiber, *Chaotic Behaviour of Deterministic Dissipative Systems,* Cambridge University Press, New York, NY, 1991.

Nonnenmacher, T.F., G.A. Losa, & E.R. Weibel, (eds.), *Fractals in Biology and Medicine,* Birkhäuser Boston, Boston, MA, 1994.

Ruelle, David, *Chance and Chaos,* Princeton University Press, Princeton, NJ, 1991.

Seydel, Rüdiger, *From Equilibrium to Chaos: Practical Bifurcation and Stability Analysis,* North-Holland, Amsterdam, 1988.

Strogatz, Steven H., *Nonlinear Dynamics and Chaos,* Addison Wesley, Reading, MA, 1994.

Szemplinska-Stupnicka, W., G. Iooss, & F.C. Moon, *Chaotic Motions in Nonlinear Dynamical Systems,* Springer-Verlag, New York, NY, 1988.

Vicsek, Tamas, *Fractal Growth Phenomena,* World Scientific, London, 1989.

Wiggins, Stephen, *Global Bifurcations and Chaos: Analytical Methods,* Springer-Verlag, New York, NY, 1988.

Zaslavsky, G.M. and L.V. Kirenskii. *Chaos in Dynamic Systems,* Harwood Academic, 1985.

Curriculur & Instructional Resources

American Association for the Advancement of Science, *Science for all Americans: Project 2061,* Oxford University Press, New York, NY, 1990.

Boyer, E. L., *Scholarship Reconsidered: Priorities of the Professoriate,* Carnegie Foundation for the Advancement of Teaching, Princeton, NJ, 1991.

Campbell, Paul J., *et al.,* (eds.), *UMAP Modules: Tools For Teaching,* Consortium for Mathematics and Its Applications, Lexington, MA, 2000.

Committee on the Undergraduate Program in Mathematics, *CUPM Curriculum Guide 2004: Undergraduate Programs and Courses in the Mathematical Science,* Mathematical Association of America, Washington, DC, 2004.

Fusaro, B.A., and P.C. Kenshaw, (eds.), *Environmental Mathematics in the Classroom,* Mathematical Association of America, Washington, DC, 2003.

Ganter, Susan and William Barker, (eds.), *Curriculum Foundations Project: Voices of the Partner Disciplines,* Mathematical Association of America, Washington, DC, 2004.

Jungck, John R., *et al.,* (eds.), *The BioQUEST Library,* Academic Press, San Diego, CA, 2002.

Jungck, John R., Ethel D. Stanley, & Marion Field Fass, (eds.), *Microbes Count! Problem Posing, Problem Solving, and Persuading Peers in Microbiology,* American Society for Microbiology Press, Washington, DC, 2003.

National Research Council, *Bio 2010: Transforming Undergraduate Education for Future Research Biologists,* National Academies Press, Washington, DC, 2003.

Seymour, Elaine and Nancy M. Hewitt, *Talking About Leaving: Why Undergraduates Leave the Sciences,* Westview Press, Boulder, CO, 1997.

Zimmermann, Walter and Steve Cunningham, (eds.), *Visualization in Teaching and Learning Mathematics,* Mathematical Association of America, Washington, DC, 1991.

Appendix I

Research and Education in Mathematics and Biology

In recent years the National Science Foundation (NSF) has awarded grants to mathematical and biological scientists both to conduct innovative research in the applications of mathematics to biology and to enhance undergraduate education at the intersection of the biological and mathematical sciences. Most research projects involve collaboration among individuals with different disciplinary backgrounds; most also require either the development of new mathematical methods or the application of established methods to new problems where their effectiveness has not been tested. Grants for undergraduate education are intended to better prepare undergraduate biology or mathematics students to pursue graduate study and careers in fields that integrate the mathematical and biological sciences.

To illustrate the nature and scope of these projects, we list below some recent awards. Although quite brief, the descriptions of research interests and titles of education awards convey the scope of biological problems that are being attacked with mathematical tools as well as the variety of mathematical methods that are being employed to solve biological problems. Additional information can be found, as noted, on the researchers' personal or laboratory web pages or on the NSF web site.

Research Projects

Sally Blower <sblower@mednet.ucla.edu>
Professor, Department of Biomathematics, University of California—Los Angeles. *Research interests*: Models of transmission dynamics. Hopes to develop the study of infectious diseases into a predictive science. *Personal Page:* www.biomath.medsch.ucla.edu/faculty/sblower

Ben Bolker <bolker@zoo.ufl.edu>
Assistant Professor, Department of Zoology, University of Florida. *Research interests*: Spatial, theoretical, mathematical, computational and statistical ecology. *Personal Page:* www.zoo.ufl.edu/bolker

Bruce Bush <bruce_bush@merck.com>
Senior Research Fellow, Merck Research Labs. *Research interests:* Bioinformatics.

Steve Carpenter <srcarpen@wisc.edu>
Halverson Professor of Limnology and Professor of Zoology, University of Wisconsin. *Research interests*: Limnology, experimental analysis of ecosystems, ecological modeling.
Personal Page: http://limnology.wisc.edu/personnel/carpenter/carpenter.html

Tom Daniel <danielt@u.washington.edu>
Joan and Richard Komen Professor of Zoology, University of Washington. *Research interests*: Biomechanics, mathematical biology; studies motion in biology, from the level of molecules to that of whole animals. *Lab page*: http://faculty.washington.edu/danielt/

Joe Felsenstein <joe@gs.washington.edu>
Professor of Genome Sciences, University of Washington. *Research interests:* Estimating population parameters from population samples of molecular sequences using computationally intensive methods

(e.g., Markov Chain Monte Carlo Integration). *Lab Page:* http://evolution.genetics.washington.edu. *Personal Page:* www.gs.washington.edu/faculty/felsenstein.htm

Richard Gomulkiewicz <gomulki@wsu.edu>

Professor, Department of Mathematics and School of Biological Sciences, Washington State University. *Research interests*: Theoretical population genetics and evolutionary biology using mathematics; focuses on how genetics and ecology interact to determine evolutionary responses of organisms to environments. *Personal Page*: www.wsu.edu/~gomulki/

Alan Hastings <sischwartz@ucdavis.edu>

Professor, Department of Environmental Science and Policy, University of California–Davis. *Research interests*: Metapopulation dynamics, fitting models to data in ecology; spatial dynamics in ecology, hybrid zone dynamics. *Personal Page*: www.des.ucdavis.edu/directory/display.asp?id=15

James Keener <keener@math.utah.edu>

Professor of Mathematics, Adjunct Professor of Bioengineering, University of Utah. *Research Interests:* Mathematical cardiology, cardiac arrhythmias; collaborates with Sasha Panfilov (U. of Utrecht) in bioinformatics; Co-author with James Sneyd (Mathematics, Aukland University) of *Mathematical Physiology. Personal Page:* www.math.utah.edu/~keener/

Mark Kirkpatrick <Kirkp@mail.utexas.edu>

Professor, Section of Integrative Biology, University of Texas. *Research interests:* Evolutionary theory, population and quantitative genetics; participant in the Center for Computational Biology and Bioinformatics which supports those interested in computational approaches to solving biological problems. *Personal Page:* www.biosci.utexas.edu/ib/faculty/ kirkpatr.htm

Nancy Kopell <nk@bu.edu>

Co-director, Center for BioDynamics; Boston University. *Research interests*: Dynamics of nervous system (especially rhythmic behavior), geometric theory of singularly perturbed systems, central pattern generators. *Personal Page:* http://cbd.bu.edu/members/nkopell.html

David McLaughlin <david.mclaughlin@nyu.edu>

Provost and Professor of Mathematics and Neural Sciences, New York University. *Research interests*: Visual neural science, integrable waves, chaotic nonlinear waves, mathematical nonlinear optics, mathematical physiology. *Personal Page*: www.math.nyu.edu/faculty/dmac/index.html

George Oster <goster@nature.berkeley.edu>

Professor, Department of Environmental Science, Policy and Management and Department of Molecular and Cellular Biology, University of California–Berkeley. *Research interests*: Construction and testing of theoretical models of molecular, cellular and developmental processes; collaborates with Hongyun Wang (Applied Mathematics and Statistics, Berkeley). *Personal Page:* www.cnr.berkeley.edu/~goster/home.html

Margaret Palmer <mp3@umail.umd.edu>

Professor, Department of Biology, University of Maryland. *Research Interests:* Examining the role of habitat fragmentation and habitat quality in the population dynamics of stream invertebrates using mathematical modeling. *Personal Page:* www.leopold.orst.edu/fellows/palmer

Charles Peskin <peskin@cims.nyu.edu>

Professor, Mathematics Department, Courant Institute of Mathematical Sciences, New York University. *Research Interests*: Modeling and simulation in medicine and life sciences. *Personal Page*: www.math.nyu.edu/faculty/peskin/index.html

Pavel Pevzner <ppevzner@cs.ucsd.edu>

Ronald R. Taylor Professor of Computer Science, University of California—San Diego. *Research interests*: Computational molecular biology, bioinformatics; fragment assembly in DNA sequencing, pattern discovery and regulatory genomics; computational mass-spectrometry, genome rearrangements. *Personal Page:* www.cs.ucsd.edu/users/ppevzner/research.html

John Rinzel <rinzel@cns.nyu.edu>

Professor, Center for Neural Science, New York University. *Research interests*: Biophysical mechanisms and theoretical foundations of neural computations; worked in mathematics research branch of NIH for 25 years. *Personal Page*: http://cns.nyu.edu/corefaculty/Rinzel.html

Stan Sawyer <sawyer@math.wustl.edu>

Professor of Mathematics, Genetics, and Biostatistics, Washington University–Saint Louis. *Research interests:* Mathematical genetics, probability, statistics. *Personal Page:* www.math.wustl.edu/~sawyer/index.html

Tamar Schlick <schlick@nyu.edu>

Professor, Departments of Mathematics and Chemistry; Courant Institute of Mathematical Sciences, New York University. *Research interests:* Mathematical biology, numerical analysis, computational chemistry. *Project Page:* http://monod.biomath.nyu.edu/

David Searls <David_B_Searls@sbphrd.com>

Vice President, SmithKline Beecham and Director, Bioinformatics, University of Pennsylvania. *Website:* www.bioinformatics.com

Ruth Shaw <shaw@superb.ecology.umn.edu>

Professor, Department of Ecology, Evolution and Behavior, University of Minnesota. *Research interests*: Evolutionary quantitative genetics, plant population biology. *Personal Page*: http://biosci.cbs.umn.edu/eeb/faculty/ ShawRuth.html

Steven Strogatz <shs7@cornell.edu>

Professor, Department of Theoretical and Applied Mathematics, Cornell University. *Research interests*: Mathematical biology, nonlinear dynamics and chaos applied to physics, engineering and biology; parametric resonance in MEMS; nonlinear dynamics of HIV interacting with the immune system. *Personal Page*: www.tam.cornell.edu/Strogatz.html

DeWitt Sumners <sumners@math.fsu.edu>

Professor and Chair, Department of Mathematics, Florida State University. *Research interests:* Biomedical mathematics (mathematical analysis of the human brain with functional data); applications of topology to molecular biology and polymer configuration. *Personal Page:* www.math.fsu.edu/~sumners/

John Tyson <tyson@vt.edu>

Professor, Biology Department, Virginia Tech. *Research Interests*: Eukaryotic cell cycle control mechanisms, wave propagation in excitable media; collaborates with Layne Watson (mathematics), Jill Sible (molecular and cell Biology), Clifford Shaffer and Naren Ramakrishnan (computer science). *Computational cell biology*: leibniz.biol.vt.edu *Personal Page:* www.biol.vt.edu/faculty/tyson/tysonhome.htm

Hongyun Wang <hongwang@ams.ucsc.edu>

Assistant Professor, Department of Applied Mathematics and Statistics, University of California–Berkeley. *Research interests:* Molecular structural analysis, animation and visualization, biophysics and molecular biology. *Personal Page:* www.cse.ucsc.edu/~hongwang/

Michael Waterman <msw@hto.usc.edu>

Professor, Departments of Biological Sciences, Mathematics and Computer Science, University of Southern California. *Research interests:* Computational biology, concentrating on creation and application of mathematics, statistics, and computer science to molecular biology, particularly to DNA, RNA and protein sequence data. *Personal Page:* http://htoe.usc.edu/people/Waterman.html

Education Projects

Examples of awards made under NSF's program for "Interdisciplinary Training for Undergraduates in Biological and Mathematical Sciences."

Azmy Ackleh & Jacoby Carter, University of Louisiana, Lafayette. *Training Undergraduate Students in Mathematical Biology* (UBM-0337506).

Alexander Badyaev, University of Arizona. *Evolution of Maternal Effects in a Model Avian System* (UBM-0337184).

William Boecklen, New Mexico State University. *An Interdisciplinary Program in Mathematical Biology* (UBM-0337789).

Amitabha Bose & Jorge Golowasch, New Jersey Institute of Technology. *An Undergraduate Biology and Mathematics Training Program at NJIT* (UBM-0436244).

Suncica Canic, University of Houston. *Nonlinear Waves in One-Dimensional and Multi-Dimensional Conser-vation Laws* (UBM-0337355).

Vincent Cassone, Texas A&M Research Foundation. *Integrated Research Experiences in Biological and Mathematical Sciences* (UBM-0436308).

Edward Connor, San Francisco State University. *Integrating Inquiry-Based and Cooperative Learning with Expanded Career Horizons in Environmental Biology* (UBM-0337803).

Edward Connor, San Francisco State. *Undergraduate Training in Quantitative Environmental Biology* (UBM-0436313).

George C. Cosner & Robert S. Cantrell, University of Miami. *Ecological Modelings: From Individual Utilization of Space to Community Structure* (UBM-0336812).

Daniel Cristol & John Swaddle, College of William and Mary. *Undergraduate Research in Metapopulation Ecology* (UBM-0436318).

Eric Delson, CUNY Graduate School University Center. *Research and Training in Evolutionary Primatology* (UBM-0337561).

Michael Dorcas, Davidson College. *Developing Student Scientists: Collaborative Research in the Life Sciences* (UBM-0336919).

Donald A. Drew, Rensselaer Polytech Institute. *Mathematical Models for Processes in Bacterial Cell Division* (UBM-0337659).

Richard Gomulkiewicz, Washington State University. *Collaborative Research: Maintaining High Species Diversity in Communities* (UBM-0337582).

Louis Gross & Suzanne Lenhart, University of Tennessee Knoxville. *Spatial Models for Invasion Biology* (UBM-0337593).

Anthony Ives, University of Wisconsin-Madison. *Undergraduate Collaborative Group in Mathematical Biology* (UBM-0337500).

Wendy Katkin, SUNY at Stony Brook. *Conference on Undergraduate Research and Scholarship and the Mission of the Research University* (UBM-0337532).

Jeff Knisley, Edith Seier, Robert Price, East Tennessee State University. *A Multi-Stage, Technology Intensive Approach to Statistics Instruction* (UBM-0337406).

Yang Kuang, James Elser, & William Fagan, Arizona State University. *Theoretical Frameworks for Ecological Dynamics Subject to Stoichiometric Constraints* (UBM-0337157).

Yang Kuang & John Anderies, Arizona State University. *Interdisciplinary Training for Under-graduates in Biological and Mathematical Sciences at ASU* (UBM-0436341).

Anthony Macula, SUNY at Geneseo. *Analysis of Biological Networks: A Symbiotic Relationship Between Information Science and Biology* (UBM-0337284).

Anthony Macula & Wendy Pogozelski, SUNY College at Geneseo. *RUI: Undergraduate Biomathematical Research Career Initiative at SUNY-Geneseo* (UBM-0436298).

Roy Mathias, College of William and Mary. *Matrix Analysis in Engineering and Science* (UBM-0337529).

Jason Miller, Alan Garvey, Dana Vazzana, Pamela Reich, & Carol Hoferkamp, Truman State University. *Mathematical Biology Initiative* (UBM-0337769).

Jason Miller & Michael Kelrick, Truman State Univer-sity. *RUI: Research-Focused Learning Communities in Mathematical Biology* (UBM-0436348).

Claudia Neuhauser, University of Minnesota - Twin Cities. *Stochastic Processes in Ecology and Population Genetics* (UBM-0336916).

J. Rojas, Texas A&M University. *Real Solving and Protein Structures* (UBM-0337752).

Timothy Seastedt, University of Colorado at Boulder. *The Niwot Ridge LTER Research Program 1998– 2004: Controls on the Structure, Function & Interactions of Alpine and Subalpine Ecosystems of the Colorado Front Range* (UBM-0337720).

Thomas Smotzer, James Mike, Gina McHenry, Scott Martin, & Robert Kramer, Youngstown State University. *Interdisciplinary Undergraduate Educa-tion Through Intensive Research Experiences in Mathematics and Biology* (UBM-0337558).

Lori Stevens & Daniel Bentil, University of Vermont & State Agricultural College. *Research Based Interdisciplinary Training for Mathematics and Biology Majors* (UBM-0436330).

Tina Straley, Mathematical Association of America. *Meeting the Challenges in Emerging Areas: Edu-cation Across the Life, Mathematical, and Computer Sciences* (UBM-0337646).

Richard Veit, CUNY Research Foundation. *CAREER: Dynamics of Predator-Prey Behavior in the Antarctic Ocean* (UBM-0337648).

Appendix 2

Web Resources for Bioinformatics, Computational Science, Mathematical Biology, and Scientific Visualization

The rapid growth in applications of quantitative and mathematical methods in the biological sciences is due in large measure to the development of the Internet and the subsequent ease with which modules, projects, software, courses, and workshops can be shared, critiqued, and improved. Consequently, the scope of web resources relevant to quantitative biology is enormous and growing, literally, exponentially. The web resources listed in this appendix represent a sample of the corpus of relevant web material. Many are sites that have been mentioned by authors of papers in this volume; others are sites that contain important links to other sites. We intend no judgment that these represent the best of their category, only that that they provide a fair sample of current web resources.

Contents

Undergraduate Programs and Courses

Majors, Minors, Programs

Bioinformatics Across the Curriculum, *University of Wisconsin System*
http://bioweb.uwlax.edu/GenWeb/Molecular/bioinfo%20curric.htm

Biotechnology Laboratory Technician Program, *Madison Area Technical College*
http://www.wcer.wisc.edu/nise/cl1/ilt/case/madison/madison.htm

Computational Science Across the Curriculum, *Capital University*
http://www.capital.edu/acad/as/csac

Computational Science Minor, *Capital University*
http://www.capital.edu/prosp/ug/computational-science.html

Coordinated Science Program, *University of British Columbia*
http://www.science.ubc.ca/~csp/

Mathematical Biology Major, *Harvey Mudd College*
http://www2.hmc.edu/www_common/biology/academics/biomath.html

QELP: Quantitative Environmental Learning Project, *Seattle Central Community College*
http://seattlecentral.edu/qelp

Undergraduate Genomics Research Project, *Wheaton College (MA)*
http://genomics.wheatoncollege.edu

Calculus for Biology Students

BioCalc: Mathematica-based Calculus for Biology Students, *University of Illinois at Urbana-Champaign*
http://www.wcer.wisc.edu/nise/cl1/ilt/case/uiuc/uiuc.htm

BioCalculus with Mathematica, *Florida State University*
http://www.math.fsu.edu/~mesterto/biocalc.html

Calculus for Biology Majors, *Appalachian State University*
http://www1.appstate.edu/~marland/classes/biocalc/biocalc.htm

Calculus for Biology I, *San Diego State University*
http://www-rohan.sdsu.edu/~jmahaffy/courses/s04/math121/index.html

Calculus for Biology II, *San Diego State University*
http://www-rohan.sdsu.edu/~jmahaffy/courses/s04/math122/index.html

Differential Calculus for Life Science Students, *University of British Columbia*
http://ugrad.math.ubc.ca/coursedoc/math102/

Integral Calculus for Life Science Students, *University of British Columbia*
http://ugrad.math.ubc.ca/coursedoc/math103/

Project CALC, *Duke University*
http://www.math.duke.edu/education/proj_calc/

The Post CALC Project, *Duke University*
http://www.math.duke.edu/education/postcalc/

Bioinformatics, Computational Science, General Education

Algorithms and DNA, *Wheaton College (MA)*
http://cs.wheatonma.edu/mleblanc/215/

Computational Biology Seminar, *Davidson College*
http://www.bio.davidson.edu/courses/compbio/webpage/home.htm

Computational Biology, *Capital University*
http://www.capital.edu/acad/as/csac/Comp_Bio/compbio.htm

Computational Science, *Capital University*
http://www.capital.edu/acad/as/csac/Comp_Sci1/compsci1.htm

DNA (A Joint Biology & Computer Science Course), *Wheaton College (MA)*
http://cs.wheatoncollege.edu/mleblanc/dna

Global Change: An Interdisciplinary Learning Environment, *University of Michigan*
http://www.wcer.wisc.edu/nise/cl1/ilt/case/michigan/michigan.htm

Integrated Science Program, *University of British Columbia*
http://www.science.ubc.ca/~isp/

Mathematical Biology, *Harvey Mudd College*
http://www.math.hmc.edu/~depillis/MATHBIO/index.html

Science One, *University of British Columbia*
http://www.science.ubc.ca/~science1/

Six Billion People and Counting, *Seattle Central Community College*
http://seattlecentral.edu/faculty/jhull/6bill.html

Understanding Our World, *James Madison University*
http://www.isat.jmu.edu/users/klevicca/idls/scicore.htm

Teaching Resources

Modules & Case Studies

A Breast Cancer Case Study
http://www.genetichealth.com/BROV_A_Case_Study.shtml

A Delicate Balance: Signs of Change in the Tropics
http://earthobservatory.nasa.gov/Study/DelicateBalance

As the HIV Virus Incubates
http://www.wolfram.com/products/mathematica/usersanduses/experience/as.html

Cell Physiology Bioinformatics Grant Exercise
http://www.wooster.edu/biology/dfraga/BIO_305/305_grant_exercise.html

Gene Identification
http://www.capital.edu/acad/as/csac/Keck/modules.html#gene

Kermack-McKendrick SIR Model for Epidemics
http://mathworld.wolfram.com/Kermack-McKendrickModel.html

Modeling the Cardiovascular System using Stella
http://www.capital.edu/acad/as/csac/Keck/modules.html#cardio

Modeling Malaria
http://www.capital.edu/acad/as/csac/Keck/modules.html#malaria

Models of Species interaction
http://www.math.duke.edu/education/postcalc/predprey/index.html

Modules for Scientific Computation and Visualization
http://www.capital.edu/acad/as/csac/Keck/modules.html

Ninos Desaparecidos Case Study: Commentary
http://ublib.buffalo.edu/libraries/projects/cases/ninos/ninos_notes.html

Ninos Desaparecidos: Genetics and Human Rights
http://ublib.buffalo.edu/libraries/projects/cases/ninos/ninos

Pulsing Polyps in Complex Colonies
http://www.wolfram.com/products/mathematica/usersanduses/experience/surna.html

SIR Model for Spread of Disease
http://www.math.duke.edu/education/ccp/materials/diffcalc/sir/sir2.html

Watching Plants Dance to the Rhythms of the Oceans
http://earthobservatory.nasa.gov/Study/SSTNDVI

Sources of Modules & Case Studies

Biology Modules from the Connected Curriculum Project
http://www.math.duke.edu/education/ccp/materials/biology.html

CCP: Connected Curriculum Project
http://www.math.duke.edu/education/ccp/

Chem Connections Workbooks
http://wwnorton.com/college/chemistry/chemconnections/

ChemConnections: Systemic Change Initiatives in Chemistry
http://chemlinks.beloit.edu

COMAP: Consortium for Mathematics and Its Applications
http://www.comap.com

MERLOT Biology: Peer Reviewed Links to Online Learning Materials
http://www.merlot.org/Home.po?discipline=Biology

National Center for Case Study Teaching in Science
http://ublib.buffalo.edu/libraries/projects/cases/case.html

Other Classroom Resources

Beyond Bio 101: Transforming Undergraduate Biology
http://www.hhmi.org/BeyondBio101/index.htm

Bio-Rad Life Sciences Education
http://www.bio-rad.com/

BugScope: Remote Access to Scanning Electron Microscope
http://bugscope.beckman.uiuc.edu

Cells Alive! A Gallery of Images and Movies
http://www.cellsalive.com/

Computational Cell Biology
http://www.compcell.appstate.edu/

Computational Sciences
http://www.carleton.ca/natsci/compsci/whatis.html

Earth Observatory: Images and Data from NASA Missions
http://earthobservatory.nasa.gov

Genetics Glossary
http://helios.bto.ed.ac.uk/bto/glossary/

Human Genome Project Education Resources
http://www.ornl.gov/sci/techresources/Human_Genome/education/education.shtml

Introduction to Stereology
http://www.liv.ac.uk/fetoxpath/quantoxpath/stereol.htm

Microbes Count: Multimedia Resources for Microbiology
http://www.bioquest.org/microbescount/

Misconceptions in Mathematics
http://www.counton.org/resources/misconceptions/index.shtml

NABT Web Resources for Biology Teachers
http://www.nabt.org/sup/resources/

Quantitative Curriculum for Life Science Students
http://www.tiem.utk.edu/~gross/quant.lifesci.html

Stereology
http://filebox.vt.edu/users/rboehrin/

The Image Processing and Measurement Cookbook
http://reindeergraphics.com/tutorial/index.shtml

Using Data in the Classroom
http://serc.carleton.edu/usingdata/index.html

VISM: Visualization in Science and Mathematics
http://www.isat.jmu.edu/common/projects/VISM/

Workshop Biology
http://yucca.uoregon.edu/wb/

Workshop Physics
http://physics.dickinson.edu/~wp_web/WP_homepage.html

Software

Comprehensive Mathematical Packages

Maple: A System for Mathematical and Scientific Computation
http://www.maplesoft.com/products/maple/

Mathematica: Integrated Environment for Scientific Computing
http://www.wolfram.com

MatLab: Technical Computing Environment
http://www.mathworks.com/products/matlab/

SciLab: A Free Open-Source Scientific Software Package
http://www.scilab.org

Visualization Software

GeoWall: Virtual Reality Visualization Devices
http://geowall.geo.lsa.umich.edu/intro.html

Graphic Analysis: Producing Graphs from Data
http://www.vernier.com/soft/ga.html

Image J: Public Domain Java Image Processing Program
http://rsb.info.nih.gov/ij/

Image Processing Toolbox
http://www.mathworks.com/products/image/

IPTK: Image Processing Tool Kit
http://www.reindeergraphics.com/iptk/

IRIS Explorer: Customized Visualization Environment
http://www.nag.co.uk/Welcome_IEC.html

NCG: NCAR Graphics Package for Scientific Visualization
http://ngwww.ucar.edu/ng/index.html

NIH Image: A Public Domain Image Processing Program
http://rsb.info.nih.gov/nih-image

Scion Image for Windows
http://www.scioncorp.com/frames/fr_scion_products.htm

Bioinformatics Packages

Bioinformatics Toolbox
http://www.mathworks.com/products/bioinfo/

BLAST: Basic Local Alignment Search Tool
http://www.ncbi.nlm.nih.gov/Education/BLASTinfo/information3.html

Genetic Algorithm and Direct Search Toolbox
http://www.mathworks.com/products/gads/

Molecular Visualization

Insight II: Molecular Modeling Environment
http://www.accelrys.com/insight/index.html

Marvin: Java Applets for Visualizing Molecules
http://www.chemaxon.com/marvin/

Molecular Visualization Freeware: Protein Explorer, Chime, RasMol
http://www.umass.edu/microbio/rasmol

MoluCAD: A Molecular Modeling and Visualization Tool
http://www.kinematics.com/molucad/

Quanta: Modeling Environment for Organic Macromolecules
http://www.accelrys.com/quanta/index.html

Raster 3D: Raster Images of Proteins or Other Molecules
http://www.bmsc.washington.edu/raster3d/raster3d.html

Geographic Information Systems (GIS):

ArcGIS: Geographic Information System (GIS) Package
http://www.esri.com/software/arcgis/index.html

My World: Geographic Information System Software
http://www.worldwatcher.northwestern.edu/myworld/index.html

Modelling Software

AUTO: Continuation and Bifurcation Problems in ODE
http://indy.cs.concordia.ca/auto/

Berkeley Madonna: Modeling and Analysis of Dynamic Systems
http://www.berkeleymadonna.com/

Stella: Simulation and Modelling Software
http://www.iseesystems.com

XPP: X-Windows Phase Plane: Solving Differential Equations
http://www.math.pitt.edu/~bard/xpp/xpp.html

Statistical and Data Analysis

IDL: Interactive Data Language
http://www.rsinc.com/idl/index.asp

NCL: NCAR Command Language for Analyzing Data
http://ngwww.ucar.edu/ncl/index.html

R: Open Source Environment for Statistical Computing
http://www.r-project.org

Inventories of Software Packages

Biology Education Software FAQs
http://www.snarkware.net/bioedusoft/

Links to Molecular Display & Visualization Software
http://www.netsci.org/Resources/Software/Modeling/Viewers

Visualization Software Links
http://serc.carleton.edu/introgeo/models/visual/software.html

Professional Development

Consortiums & Collaboratives

BioQuest Curriculum Consortium, *Beloit College*
http://www.bioquest.org

C*ODE*E: Consortium of ODE Experiments, *Harvey Mudd College*
http://www.math.hmc.edu/codee/main.html

CBE: Center for Biology Education, *Universityof Wisconsin-Madison*
http://www.wisc.edu/cbe/index.html

Center for Quantitative Life Sciences, *Harvey Mudd College*
http://www.math.hmc.edu/~depillis/KECK_QLS/index1.html

ChemConnections, *University of California-Berkeley*
http://mc2.cchem.berkeley.edu/

Collaborative Web Site for Biology Education, *University of Wisconsin System*
http://bioweb.uwlax.edu/

Keck Undergraduate Computational Science Education Consortium, *Capital University*
http://www.capital.edu/acad/as/csac/Keck

Workshops

Biotechnology Summer Institute: Bio-techniques in Teaching, *J. Sargeant Reynolds Community College*
http://staff.jsr.cc.va.us/asullivan/eisenhower/

Chautauqua Short Courses for College Teachers
http://www.chautauqua.pitt.edu/index.html

NCSI: National Computational Science Institute
http://www.computationalscience.org

PKal: Project Kaleidoscope
http://www.pkal.org

PREP: Professional Enhancement Programs, *Mathematical Association of America*
http://www.maa.org/prep/

Summer Institute on Undergraduate Education in Biology, *National Academies of Science*
http://www.academiessummerinstitute.org/

Reports

Bio 2010: Transforming Undergraduate Education
http://www.nap.edu/books/0309085357/html/

Mathematics in Biology (Science, February 2004)
http://www.sciencemag.org/sciext/mathbio/

Mathematics Curriculum Foundations for Biology
http://www.maa.org/cupm/crafty/focus/cf_biology.html

Mathematics and Biology: The Interface
http://www.bio.vu.nl/nvtb/Contents.html

Research Programs

Biocomputing Laboratory, *University of Washington*
http://depts.washington.edu/biocomp/

BioMaps: Institute for Quantitative Biology, *Rutgers University*
http://www.biomaps.rutgers.edu/index.html

Brutlag Bioinformatics Group, *Stanford University*
http://motif.stanford.edu/

Computational and Genomic Biology, *University of California-Berkeley*
http://computationalbiology.berkeley.edu/

Mathematical Biology, *Brandeis University*
http://www.bio.brandeis.edu/biomath/menu.html

MBI: Mathematical Biosciences Institute, *Ohio State University*
http://mbi.osu.edu

UCSD Bioinformatics Laboratory, *University of California-San Diego*
http://scilib.ucsd.edu/bioinf_lab/

Societies, Institutions, Portals

Professional Societies

AAAS: American Association for the Advancement of Science
http://www.aaas.org

ASM: American Society for Microbiology
http://www.asm.org

Ecological Society of America
http://www.esa.org/

ISCB: International Society for Computational Biology
http://www.iscb.org/

ISS: International Society for Stereology
http://www.stereologysociety.org/

MAA: Mathematical Association of America
http://www.maa.org

NABT: National Association of Biology Teachers
http://www.nabt.org

NSTA: National Science Teachers Association
http://www.nsta.org

SIGSCE: Special Interest Group on Computer Science Education
http://www.sigcse.org

SMB: Society for Mathematical Biology
http://www.smb.org

Institutions

Complex Biological Systems Initiative
http://www.nigms.nih.gov/funding/complex_systems.html

ESRI: Environmental Systems Research Institute
http://www.esri.com/index.html

HHMI: Howard Hughes Medical Institute
http://www.hhmi.org

Institute for Environmental Modeling
http://www.tiem.utk.edu/

NIGMS: National Institute of General Medical Sciences
http://www.nigms.nih.gov

NSF: National Science Foundation
http://www.nsf.gov

Web Portals (Biology and Biotechnology)

Access Excellence: The National Health Museum
http://www.accessexcellence.org

BEN: BioScience Education Network (BioSciEdNet) (BEN)
http://www.biosciednet.org

The Virtual Library: BioSciences
http://vlib.org/Biosciences.html

Educating the Biotechnology Work Force
http://www.bio-link.org/

Biotechnology Information Directory
http://www.cato.com/biotech/bio-edu.html

NCBI: National Center for Biotechnology Information
http://www.ncbi.nlm.nih.gov

Computational Molecular Biology at NIH
http://molbio.info.nih.gov/molbio/

Pedro's BioMolecular Research Tools
http://www.public.iastate.edu/~pedro/research_tools.html

Meeting the Challenges

http://www.maa.org/mtc/

NSDL: National Science Digital Library
http://www.nsdl.org

Web Portals (Geomics and Bioinformatics)

Bioinformatics at the NIH
http://www.bisti.nih.gov/

Genetics Education Center
http://www.kumc.edu/gec/

Genomics GTL: Using Genomic Tools for Proteinomics
http://doegenomestolife.org/

Genomics: From DNA to Life
http://www.doegenomes.org

Human Genome Project Information
http://www.ornl.gov/sci/techresources/Human_Genome/home.shtml

Microbial Genome Project
http://www.ornl.gov/sci/microbialgenomes/

BioCatalogue (European Bioinformatics Institute)
http://www.ebi.ac.uk/biocat/biocat.html

International List of Bioinformatics Courses
http://www.nslij-genetics.org/bioinfotraining/

Web Portals (Mathematical Biology)

Internet Mathematics Library: Biology Applications
http://mathforum.org/library/topics/biology/branch.html

Math Forum
http://mathforum.org

Mathematical Life Sciences Archives
http://archives.math.utk.edu/mathbio/

Resource Page for Mathematical Biology
http://www1.appstate.edu/~marland/

Appendix 3

COMAP Modules

The Consortium for Mathematics and its Applications (COMAP) offers an extensive set of modules focused on using mathematics to model real-world systems. The following are abstracts for many of the biological modules. Additional information is available at www.comap.com.

Population Models in Biology and Demography (#99777)

This module presents applications from microbiology and demography that give a physical context for critical concepts in a first-semester calculus course. Elementary models of population growth are developed using data collected from biological experiments conducted during the course. Then, proceeding from "microcosm to macrocosm," the tools of calculus, demographic software, and the Maple computer algebra system (CAS) are used to address questions of human population projections. Included are discussion materials, sample models using exponential and logistic curves, projects, complete biology labs for generating raw data, exercises for developing facility with Maple, and resource lists (including software and Internet sites) for studying demographic questions.

Small Mammal Dispersion (#99776)

This module introduces students to the social fence hypothesis explaining small mammal migration between adjacent land areas. Students are shown how the hypothesis is formulated in the population ecology literature as a pair of autonomous differential equations, and then they are directed toward a modified version of the standard formulation leading to increased realism. The modified version is solved qualitatively with phase diagrams for a range of ecological circumstances. Students also gain experience working with the numerical phaseplane plotter Dynasys, which can be downloaded from the World Wide Web. The social fence hypothesis is presented within the real-world context of controlling beaver-related damage in a given area by trapping.

The Resilience of Grassland Ecosystems (#99775)

This module introduces students to the state-and transition theory explaining the succession of plant species on grassland and to the concept of successional thresholds partitioning plant states into those gravitating toward socially desirable or socially undesirable plant compositions over time. Students are shown how the state-and-transition theory is formulated in the mathematical ecology literature as a system of two autonomous differential equations, and how a successional threshold is defined by the stable manifold to an interior saddle-point equilibrium. A series of exercises directs students toward a qualitative phase-plane solution of the system and an analytical approximation of the stable manifold. Students also gain experience working with the numerical phase-plane plotter Dynasys, which can be downloaded from the web. A discussion section applies the approximated stable manifold to the real-world problem of controlling livestock numbers on public grazing land to reestablish more socially-desirable plant varieties. The module is within the capabilities of students having had basic calculus and an introductory course in ordinary differential equations covering phase-plane solutions.

The Species — Area Relations (#99768)

This unit examines the fundamental ecological relationship that as the area A of a region increases, so does the number of different species S encountered. We begin by motivating why power curves $S = cAz$ are often used to model this relationship. Using logarithms, we show how a model curve can be fit to species–area data sets. We present various interpretations and uses of the constants c and z, and show how the model has been used by researchers to describe species loss resulting from deforestation. Real data sets are used extensively throughout the unit. Descriptions of student field projects are included.

Of Mites and Models: A Temperature-Dependant Model of a Mite Predator-Prey Interaction (#99764)

We analyze the qualitative behavior of a model for a mite predator-prey interaction. This model is based on a simple system of differential equations, and the model parameters are assigned values determined for a specific interaction between two species of mites. Several of these parameters are functions of temperature, and temperature is treated as a bifurcation parameter in the analysis of the model. It is shown that, depending on the temperature value, the model exhibits a stable fixed point, a stable limit cycle, or both (bistability). The model is used to illustrate population outbreaks.

Optimal Foraging Theory (#99762)

This unit is an introduction to the modeling of animal foraging behavior. We look first at how such basic factors as search and handling times and energy content may affect prey choice by foraging animals. We construct an elementary model that maximizes the net energy intake rate of the forager and that can predict "optimal" diets of animals. We then modify and refine this model to take into account factors such as prey recognition time, food patches, and central place foraging. Data from several experiments illustrate all the models.

The Hardy-Weinberg Equilibrium (#99738)

Using elementary probability, this unit shows how genotypes in a population reach an equilibrium in a single generation, under appropriate conditions. Real data illustrate this concept, and conditions are examined that may keep a population from attaining an equilibrium.

Information Theory and Biological Diversity (#99705)

Discusses and derives the key properties of one measure of diversity, the entropy function, and illustrates its use by ecologists and animal behaviorists.

Time Resources in Animals (#99688)

This unit presents an alternative to the classical optimal foraging models in behavioral ecology. In contrast to optimizing the net energy intake in a forager's diet, the model presented in this reading is concerned with a time-budgeting process dependent only upon whether an animal is hungry or satiated at a given moment. The analysis of the model is carried out using simple Markov chains. Computer programs included in the unit are used to generate "field data," which are then used to determine the proportion of time various animals will spend foraging and resting. This process exposes beginning students to elementary analysis of rather complex data sets.

Biokinetics of a Radioactive Tracer (#99686)

A system of differential equations is used to describe the transfer and breakdown of I_{131} aluminum in rabbits. The original research was by E.B. Reeve and J.E. Roberts.

Compartment Models in Biology (#99676)

This module introduces compartment models and their applications.

The Lotka-Voltera Predator-Prey Model (#99675)

This module describes and analyzes qualitatively a simplified version of the predator-prey model attributed to Lotka and Volterra. Deductions are made concerning the size of populations based on information about their percentage growth rates. The module describes a non-standard and stimulating way of illustrating the power and utility of combining geometry and algebra.

Least-Squares, Fish Ecology and the Chain Rule (#99670)

Differential equation models are used to study the effect of the amount and spatial distribution of the food supply on the amount fish eat. These models are combined into a single, more encompassing model via the multivariable chain rule. The original research was by V.S. Ivlev, a Soviet biologist. His experimental data is included. The use of least-squares techniques to fit model parameters to the data is emphasized and is discussed in detail in an appendix that may be used independently as an introduction to the concepts and practice of the least-squares method.

Differential Growth, Huxley's Alometric Formula, and Sigmoid Growth (#99635)

Differential growth refers to the changes in the proportions, morphological or chemical, following the increase in size, inside a single species or between adults of related species. In many cases differential growth is controlled by the allometric formula $Y = nX^m$, establishing a correlation of growth found in many sectors of biology, from embryology, taxonomy, and paleontology to ecology and zoology. This entirely empirical formula arises, as we will see, in a variety of equivalent forms. The exercises will show that it is related to the logistic formulae of absolute growth representing S-shaped curves. The great appeal of the allometric formula comes mostly from its simplicity, from the frequency with which it arises and from its close relation to Thompson's famous Theory of Transformations. This formula is the most versatile mathematical expression for intra- or interspecies comparisons.

Determining the Size for a Mussel Culture Farm in a Tidal Estuary Based on Local Biological Factors (#99607)

In recent years, experimental mussel culture farms have been constructed along the coasts of the United States to examine the feasibility of growing mussels in large scale commercial farms. A mathematical model is provided to estimate the size of such a farm which will maximize the harvest for a given location without overly depleting nutrients in a tidal estuary. Students learn: 1) to better understand how simple mathematical models can be used to predict results; 2) to recognize the importance of preliminary research data employed in a mathematical model; 3) to recognize the need for making idealized assumptions and examining ways in which certain assumptions can be altered; and 4) to gain practice in developing skills in computing and using proper units of measurement.

The Use of Continued Fractions in Botany (#99571)

The use of continued fractions in botany has allowed workers to make important steps in the direction of the solution of the so-called problem of phyllotaxis. This problem is characterized by the emergence of the series 1, 1, 2, 3, 5, 8, 13, 21,... in the secondary spirals seen on plants. The question is, "Why does this series arise in 95% of the observations on plants?" It has become a major preoccupation for a lot of research workers in biomathematics. Even if there is not yet a complete answer to that question, there are results, some of which will be put here into light by an accessible, original, straightforward, and pedagogical presentation. Students will become acquainted with some of the most interesting aspects of continued fractions, realize and appreciate the role of some mathematical concepts in botany, develop observational capacities in the surrounding nature, acquire the fundamental notions that will eventually allow them to investigate an important field of research in biomathematics.

Graphical Analysis of Some Difference Equations in Biology (#99553)

The growth of many biological populations can be described by difference equations. This module shows how the behavior of the solutions to certain equations can be predicted by graphical techniques. Students learn how to formulate models of population growth and use graphical and numerical techniques to analyze the behavior of populations.

The Design of Honeycombs (#99502)

Why Do Cells Compete? (#99484)

This unit introduces current data about brightness constancy and contrast in visual perception, and shows that data of this kind are a consequence of a principle that holds in all cellular systems. Variations on this principle imply other visual properties, such as edge detection, spatial frequency detection, and pattern-matching properties.

Curve Fitting Via the Criterion of Least Squares (#99321)

Testing a Hypothesis: t-test for Independent Samples (#99268)

The presentation of a problem and its solution indicating the role of the statistical t-test of significance in the research process. The example here is biological. Students perform the t-test on experimental data, interpret the calculated value of "t", and gain an understanding of the general relationship between the statistical arithmetic and the design and conduct of experiments.

The Human Cough (#99211)

Educational objectives: 1) To see how a physical assumption may lead to a choice of domain for a function; 2) To see an application of maximization of a function on a closed interval domain; 3) To interrelate biology, physics and calculus.

Selection in Genetics (#99070)

Output skills: 1) Be able to describe how a recurrence relation for gene frequency in the nth generation is obtained; 2) Be able to explain the use of calculus to approximate the result obtained from this recurrence relation.

The Digestive Process of Sheep (#99069)

This unit introduces a differential equations model for the digestive processes of sheep. Students describe the digestive processes of ruminants, explain how the assumptions of the model are translated into equations, discuss what support there is for the validity of the model, and discuss some possible conclusions to be drawn from the model.

Population Growth and the Logistic Curve (#99068)

Output skills: 1) Be able to describe the mathematical assumptions that lead to the logistic equation; 2) Be able to criticize the logistic model and to discuss some of its strengths and weaknesses; 3) Know the behavior of the logistic function and the shape of its graph.

Modeling the Nervous System: Reaction Time and the Central Nervous System (#99067)

Output skills: 1) Be able to describe the process by which the central nervous system responds to a stimulus; 2) Be able to explain how this process is modeled mathematically and what conclusions can be drawn from the model regarding reaction time; 3) Be able to discuss the merits of various assumptions concerning the relation between intensity and excitation and stimulus intensity.